The *Daily Warm-Ups series* is a wonderful way to turn extra classroom minutes into valuable learning time. The 180 quick activities—one for each day of the school year—review, practice, and teach physics. These daily activities may be used at the very beginning of class to get students into learning mode, near the end of class to make good educational use of that transitional time, in the middle of class to shift gears between lessons—or whenever else you have minutes that now go unused. In addition to providing students with fascinating physics activities, they are a natural path to other classroom activities involving critical thinking.

Daily Warm-Ups are easy-to-use reproducibles—simply photocopy the day's activity and distribute it. Or make a transparency of the activity and project it on the board. You may want to use the activities for extra-credit points or as a check on critical-thinking skills and problem-solving skills.

However you choose to use them, *Daily Warm-Ups* are a convenient and useful supplement to your regular lesson plans. Make every minute of your class time count!

Daily *warm-ups*

PHYSICS

J. WESTON

WALCH
PUBLISHER

Portland, Maine

1   2   3   4   5   6   7   8   9   10

ISBN 0-8251-4633-X

Copyright © 2003

J. Weston Walch, Publisher

P.O. Box 658 • Portland, Maine 04104-0658

walch.com

Printed in the United States of America

Science vs. Superstition

Break a mirror and you'll have seven years of bad luck! Both science and superstition observe events and develop possible explanations for how or why they occurred. The key difference is that only in the **scientific method** is the explanation tested to see if it is true or not.

State a common superstition and then describe an experiment that could be used to prove or disprove its validity.

1

Controversial Theories

In the fourth century B.C.E., Aristotle presented a model of the universe that had Earth at its center with everything else, including the Sun and stars, revolving around it. The model, which was later detailed by Ptolemy, actually worked quite well in explaining the motion of the sun, moon, and planets as viewed from Earth. In 1543, Nicolaus Copernicus published a model of the solar system that had Earth and the other planets revolving around the Sun. Although we now know the latter to be true, it took almost a century to overcome false assumptions and closed minds before it was generally accepted.

Briefly describe two current scientific theories that are in conflict with each other and give their main points of disagreement.

2

How Do We Know? 1

We often study scientific facts, laws, and theories as if they were always known to those before us. However, much of what we study is relatively new or revised information. It has been estimated that scientific knowledge has more than doubled in just the past few decades. Most of that knowledge is gained through **research,** which is scientific investigation and inquiry.

The famous rocket scientist Werhner von Braun once said research is "what I'm doing when I don't know what I'm doing." Write a paragraph explaining what he meant by that statement.

3

How Do We Know? II

Scientists (and science students) often use the **scientific method** when attempting to determine why something happened in the way it was observed to happen. For any event, several different theories can be offered as possible explanations. In many instances, a well-planned experiment can either prove or disprove the theory. In either case, knowledge has been advanced.

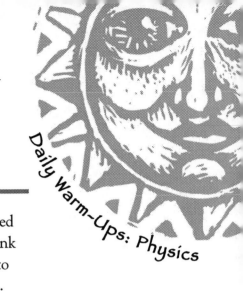

The Greek philosopher Aristotle wrote that "the mark of an educated mind is to be able to entertain a thought without accepting it." Think of a controversial scientific theory and explain why it is important to know what the theory says even if you don't agree with it.

4

How Do We Know? III

Many breakthroughs in science, such as the discovery of the nucleus, begin with a reconsideration of basic knowledge that leads to technological advances never imagined by its discoverer. Because the outcome of research is often an unknown quantity, there are times when research aimed at gaining new knowledge is controversial. Some people fear that new knowledge may lead to questionable technology.

Nobel Prize–winner Marie Curie once said, "Nothing in life is to be feared, only studied." Write a paragraph in which you explain why you agree or disagree with this statement.

Why Do We Know?

When scientific discovery leads to the invention of products useful to a society, the effects on that society are like the ripples from a stone dropped into the water. Consider for a moment how computers have revolutionized modern life.

The eighteenth-century scientist Michael Faraday made many discoveries in the fields of electricity, magnetism, and chemistry. Faraday was once asked by a minister, "What use is your discovery?" He replied, "One day you will tax it." Write a paragraph explaining what Faraday meant by his statement.

6

Daily Warm-Ups: Physics

Measure Up!

In science class, we learn the various units of measurement, such as meters (length), grams (mass), and seconds (time). Usually, these are measured in class with rulers, balances, and stopwatches. Similar measurements are made each day at home and at work with instruments that may look different from those you use in school.

For each situation below, identify a device used to make the measurement.

Situation   **Device Used**

1. a carpenter measuring a wall stud   _____

2. a cook measuring cooking time   _____

3. a chemist mixing exact amounts of ingredients   _____

4. a machinist measuring parts to exact specifications _____

7

Order in the Court

A skill that many scientists cultivate is the ability to estimate quantities to the correct **order of magnitude.** For example, Ole Roemer, a Danish astronomer, in 1675 became the first person to put the speed of light in the correct order of magnitude (10^8 m/s). This allowed experiments to be designed to measure the speed of light accurately.

Estimate the order of magnitude in meters for each of the following measurements.

1. the diameter of a tennis ball $\quad$ 10——

2. the length of a tennis racket $\quad$ 10——

3. the width of a tennis court $\quad$ 10——

4. the length of a tennis court $\quad$ 10——

5. the height of a tennis net $\quad$ 10——

8

What Size?

Units refer to how much of a given quantity is being measured. Scientists prefer to use the metric system since it is based on units of 10. However, some of the units in the English system are still very much in use in our everyday lives.

Use the list below to match the appropriate units to the quantities. Write the correct unit on the line provided. Some quantities have more than one appropriate unit in the list.

meter	*newton*	*second*	*horsepower*	*pound*
watt	*kilogram*	*ounce*	*mile*	*joule*

	Quantity	Units		Quantity	Units
1.	mass	_____	5.	power	_____
2.	distance	_____	6.	time	_____
3.	energy	_____	7.	work	_____
4.	force	_____	8.	weight	_____

9

Famous Units

Several units used in science are named after famous physicists whose contributions made them worthy of that honor.

For each of these "famous" units, give the quantity for which it is a measure and a contribution made by the physicist after whom the unit is named.

10

	Quantity	Famous Unit	Contribution
1.	_____	newton	_____
2.	_____	joule	_____
3.	_____	watt	_____
4.	_____	coulomb	_____
5.	_____	curie	_____

Unit Family Tree

All quantities that are defined in science are measured in given amounts referred to as units. Most quantities are combinations of just three basic quantities and therefore their units are combinations of three basic units. These quantities—**distance, mass,** and **time**—are called the fundamental quantities. In the mks system of measurement, their units are meters, kilograms, and seconds.

For each quantity listed below, the mks unit for that quantity is also given. In the blank to the right, give the formula in meters, kilograms, and seconds that make up that unit. *Hint:* Consider the equation that defines that quantity.

	Quantity	Unit	Formula
1.	force	newton	_____
2.	energy	joule	_____
3.	pressure	pascal	_____
4.	power	watt	_____

11

Near and Far

Most of the distances we need to know can be measured in units such as centimeters or inches, if they are small, or kilometers and miles, if they are large. But what if we are measuring objects that are extremely small or distances that are unbelievably large? Science has those covered as well.

Daily Warm-Ups: Physics

Match each unit on the right with the type of measurement with which it would most likely be used.

Measurement	Unit
_____ 1. diameter of a hydrogen atom	a. light-year (L.Y.)
_____ 2. distance from the Sun to Earth	b. nanometer (nm)
_____ 3. wavelength of a photon of visible light	c. angstrom (Å)
_____ 4. distance from our Sun to the nearest star	d. micron (μ)
_____ 5. diameter of a red blood cell	e. astronomical unit (A.U.)

12

Star Trek

Science fiction often implies that traveling to distant stars and solar systems is as easy to do as taking a cross-country trip. Of course, their spaceships travel at a speed near the speed of light or even faster! Our fastest spacecraft to date travels at a speed less than 1/100 of 1% of the speed of light.

Using the speed of the Apollo spacecraft that took astronauts to the moon (40,225 kilometers per hour), and recalling that one light-year is 9.66 trillion kilometers (9.66×10^{12}), determine the amount of time it would take today's astronauts to reach the following destinations.

	Destination	Distance from Earth	Travel Time
1.	Venus	40,234,000 kilometers	_____
2.	Mars	78,375,000 kilometers	_____
3.	Pluto	5,740,530,000 kilometers	_____
4.	Alpha Centauri	4.3 light-years	_____

13

© 2003 J. Weston Walch, Publisher

Andromeda

The **Andromeda galaxy** is the farthest object we can see in the night sky without the aid of a telescope. It is 2.25×10^{19} kilometers away from Earth.

When we look at the Andromeda galaxy, how many years are we looking into the past? *Hint:* The speed of light is 3×10^8 meters per second and one light-year is the distance that light travels in one year.

Daily Warm-Ups: Physics

14

GPS

The use of GPS has become commonplace in many of our daily activities. Drivers in automobiles use them to get directions when they are lost, and golfers use them to get the correct yardage on their next shot. Although GPS was designed for and is operated by the U.S. military, it has thousands of civil and commercial users worldwide. The system consists of a minimum of 24 satellites that orbit Earth in 12 hours. Four GPS signals are used to compute positions in three dimensions, velocity, and time.

Solve the letter tile puzzle to find out what the acronym GPS stands for.

M		S T E	B A L	G L O	S Y	I N G	S I T	I O N
P O								

15

Which Velocity Is It?

There are two types of velocity that we encounter in our everyday lives. **Instantaneous velocity** refers to how fast something is moving at a particular point in time, while **average velocity** refers to the average speed something travels over a given period of time.

For each use of velocity described below, identify whether it is instantaneous velocity or average velocity.

1. The speedometer on your car indicates you are going 65 mph. _____

2. A race-car driver was listed as driving 120 mph for the entire race. _____

3. A freely falling object has a speed of 19.6 m/s after 2 seconds of fall in a vacuum. _____

4. The speed limit sign says 45 mph. _____

Racing Against Time

The **average velocity** of an object can be found by dividing the distance an object travels by the amount of time it took to travel that distance. Sometimes positions in races are determined by the highest average speeds obtained by the cars as they qualify in individual heats.

Suppose a race-car driver needed to average 100 mph in a qualifying heat in order to gain the top spot. The heat is 100 miles. At exactly the halfway mark, her crew radios her that she has only averaged 50 mph. If the car is capable of going 150 mph, can she still make the top spot? Defend your answer.

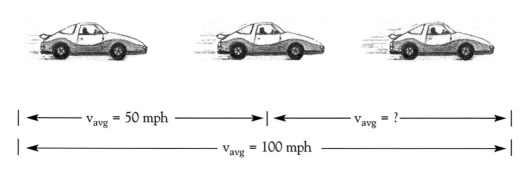

$|\longleftarrow v_{avg} = 50 \text{ mph} \longrightarrow| |\longleftarrow v_{avg} = ? \longrightarrow|$

$|\longleftarrow v_{avg} = 100 \text{ mph} \longrightarrow|$

17

Galileo the Inventor

Galileo Galilei (1564–1642) was one of those rare scientists who was both a great thinker and a great inventor. Albert Einstein took us to new levels of understanding with his thinking, but rarely spent time in a laboratory. Ernest Rutherford's expertise in experimenting led to the discovery of the nucleus, but he had trouble thinking through the entire structure of the atom. Galileo thought differently about many of the models taught in his day, and his ability to invent equipment helped him to prove many of his ideas.

```
P Z T M M M B O O F E Z
E T H E R M A S C O P E
N G I Z L M X E O J X A
D U W A T E R C L O C K
U Q R C P H S Q T K X
L M A H T O P C Q M W H
U X B N M I V O J Q P
M H K R Q P L W W P O A
U C E S Z S B Z F K E K
```

Solve the word puzzle to the left to find four of Galileo's inventions.

Movin' On

Acceleration refers to any change in an object's velocity. Velocity not only refers to an object's speed but also its direction. The direction of an object's acceleration is the same as the direction of the force causing it.

Complete the table below by drawing arrows to indicate the directions of the object's velocity and acceleration.

Description of Motion	Direction of Velocity	Direction of Acceleration
A ball is dropped from a ladder.		
A car is moving to the right when the driver applies the brakes to slow down.		
A ball tied to a string and being swung clockwise is at the top of its circular path.		
A sled is pushed to the left causing it to speed up.		

19

Show Me the Way

Quantities requiring both an amount and a direction in order to describe them are called vector quantities. When someone gives directions on how to get somewhere, they give not only distances but also the direction for each distance. In other words, they are using displacement vectors.

Using only straight lines, design a set of instructions to get from point **A** to point **B** without going through the lake. Use a scale of 1 cm = 10 m and the directional key given.

B

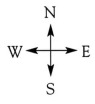

20

A

Drop and Roll

The **acceleration due to gravity (g)** near Earth's surface is 9.8 m/s^2 and is directed down toward the center of the earth. Neglecting frictional effects, any object that is dropped will accelerate in this manner. Objects that are released and then roll down an incline will accelerate with some component of *g*, depending on the angle of the incline's surface.

Use the arrow drawn to the right as the vector representation of *g*. For each of the three situations below, draw arrows showing the relative values of the ball's acceleration at each of the positions.

A ball is dropped straight down

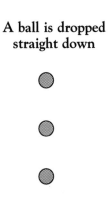

A ball rolls down a straight incline

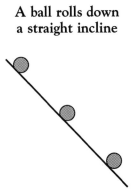

A ball rolls down a curved incline

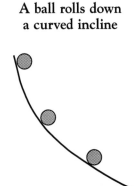

21

Newton's Laws

In the 1600s, Isaac Newton described the basis for the study of all movement with his three laws of motion. They can each be summarized with a simple question and answer.

Newton's first law of motion:

Question: What happens when no force is acting on an object?
Answer: Nothing!

Newton's second law of motion:

Question: What happens when a force does act on an object?
Answer: It accelerates in the direction of the force.

Newton's third law of motion:

Question: What happens to the object that is exerting the force?
Answer: It accelerates in the opposite direction.

Daily Warm-Ups: Physics

22

Complete the table by giving an example for each of Newton's three laws of motion.

Newton's Law of Motion	Example
First law	
Second law	
Third law	

Daily Warm-Ups: Physics

Speed of Light 1

The first attempt to measure the speed of light was made in 1630 by Galileo. He and his assistant went to the tops of nearby hills with lanterns. Galileo raised his shade to let his light escape. The instant his assistant saw the light, he in turn raised his. They repeated the experiment many times and averaged the times measured. Galileo attributed this elapsed time to his assistant's human reaction time. He reasoned that if they repeated the same experiment on far away hill tops, any difference in the time would be due to the increased distance light had to travel. When they tried it on hilltops much farther apart, they found no time difference.

A human reaction time of 0.15 seconds would be extremely fast. The speed of light is 3×10^8 meters per second. Suppose the distant hilltops used by Galileo and his assistant were 8 kilometers apart. Use calculations to explain why they found no time difference in their set of experiments.

23

Speed of Light II

Galileo's early attempt to measure the speed of light looks somewhat foolish to us now. However, he did not have the advantage of even knowing in what order of magnitude that speed would be. Oddly enough, some of Galileo's other work eventually led to that significant discovery. In 1676, Ole Roemer used a pendulum clock (built using the properties of pendulums discovered by Galileo) to accurately measure the periods of the four large moons of Jupiter (also discovered by Galileo). Roemer's calculations gave a value of 212,400 kilometers per second for the speed of light.

Daily Warm-Ups: Physics

24

Roemer found that, over the course of a year, the periods of the moons varied by as much as 16 seconds, when they should have been the same each time. He reasoned that the extra 16 seconds were due to the extra time that the light needed to travel the diameter of Earth's orbit. His value of the speed of light was off because he was using an incorrect value for the diameter of Earth's orbit. Calculate the value he was using for the Earth's orbit, and explain why getting the correct order of magnitude was so important.

Speed of Light III

The first good measurement of the speed of light was made by Albert Michelson in 1879. His work earned him the 1907 Nobel Prize. In 1927, he obtained an even more accurate measurement by duplicating Galileo's famous failed attempt some 300 years earlier. Of course, Michelson had the advantages of a precise timing device and better light sources.

Michelson's rerun of Galileo's experiment in 1927 gave a result of 299,813 kilometers per second. If the time measured for the light to make one round trip was 0.00054 seconds, how far apart were the mountain tops?

25

Flight Patterns

When an object is moving through the air and gravity is the only appreciable force (including air friction) acting on it, it is called a **projectile.**

Put a check mark by each of the following motions that are examples of projectiles. Explain why the remainder are not projectiles.

1. a football after it leaves the kicker's foot   _____

2. a golf ball in flight   _____

3. a remote control model plane in flight   _____

4. a baseball thrown by a pitcher   _____

5. a water rocket in flight   _____

26

Up, Up, and Away

A projectile launched at a 45° angle will travel farther than when launched at any other angle if the initial speed is kept the same. At this angle, the horizontal and vertical components of the projectile's velocity at launch are equal.

In 1957, Glenn Gorbus threw a regulation-sized baseball a distance of 135.9 meters. As of this writing, it remains a world-record throw. Assuming that the ball was thrown at an angle of 45° for maximum distance, calculate the speed at which it was thrown. *Hint:* Remember that at 45°, $v_h = v_v$.

27

Block That Kick

The kicking of a football is an excellent example of **projectile motion.** As the football travels along its parabolic path, its velocity has both a horizontal component, which does not change, and a vertical component, which is always changing due to gravity.

There are many reasons why more field goal attempts are missed from longer distances than from shorter distances. However, there is only one compelling reason why more field goal attempts are blocked from longer distances than from shorter distances. Use the basic physics principles of projectile motion to explain why.

28

Turn, Turn, Turn

Velocity is used to describe how fast an object is moving in various ways. **Translational velocity** refers to how fast an object moves in a straight line path. **Tangential velocity** is the speed of an object as it moves along a circular path. **Angular velocity** describes how fast an object is spinning while in rotational motion.

Identify which velocity is the correct velocity to use in each of the situations described.

Description	Velocity
1. A NASCAR race car goes around the first turn.	_____
2. Your favorite compact disc plays in a CD player.	_____
3. A child rides on a merry-go-round.	_____
4. A race horse is coming down the home stretch.	_____

29

Spinning in Space

One design for a space station in which occupants could live and work for long periods of time resembles a giant spinning wheel. The rim of the station would contain the living quarters as well as the exercise and recreational areas. Since the station spins, occupants in the rim would experience centripetal acceleration that would simulate the acceleration due to gravity on Earth.

The centripetal acceleration of an object moving in a circular path, such as the occupants in the spinning space station, can be calculated using $a_c = v^2/r$, where v is the tangential speed of the occupants and r is the radius of the rim. If the space station has a diameter of 250 m, at what rate would it have to spin in order for the occupants to experience an acceleration of 9.8 m/s²?

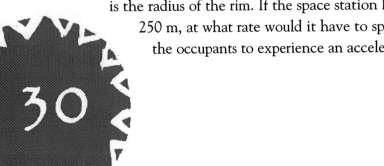

30

Round 'n' Round

When rock-and-roll music first became popular in the 1950s, records with a single song on each side were recorded at 78 revolutions per minute (rpm). Shortly thereafter, smaller records known as 45s became the standard for decades until their replacement by compact discs.

Some record collectors have valuable sets of the same song recorded on both 78-rpm and 45-rpm records. Even though they both have the same songs recorded, the diameters of the records are substantially different. Explain why.

Orbital Period

When we watch a television interview from halfway around the world, we notice a significant time delay between one journalist asking a question and the other beginning her answer. This is mainly due to the long distance the radio waves have to travel between Earth and the communications satellite. One of the many communications satellites now in orbit is 35,900 kilometers (22,300 miles) above Earth and moves with a tangential speed of 11,060 kilometers per hour (6,870 miles per hour).

Using the information given above, calculate the orbital period of the satellite. Then, comment on why your answer is significant. *Hint:* Don't forget that the diameter of Earth is 12,756 kilometers (7,926 miles).

32

Orbit Hierarchy

When the Soviet Union launched Sputnik into orbit in 1957, the race to conquer space went into high gear. Thousands of objects now orbit our planet for a wide variety of applications. We have gone from a world that thought it could not live with a satellite orbiting overhead to a world that can't imagine living without them.

Fortunately, not everything that orbits Earth does so at the same altitude. The list below contains five objects that travel, or have traveled, through space at some altitude. Not all are in orbit. Rearrange the list according to the altitude at which they travel, beginning with the lowest altitude.

Object	Order of Altitude	
Space shuttle	_____	(lowest)
Hubble space telescope	_____	
Boeing 747 airplane	_____	
Freedom 7 Mercury space capsule	_____	
International Space Station (ISS)	_____	(highest)

33

Round the World

The speed of light, 3×10^8 meters per second, is the theoretical speed limit for anything in the universe. The fastest any person has ever traveled—40,225 kilometers per hour—was on the spacecraft used to go to the Moon.

To give us a better idea of how much faster light travels than spacecraft, let's look at a couple of examples.

- The Moon is 380,000 kilometers from Earth. Calculate how long it takes light to travel from the Moon to Earth. Compare that to how long it would take the spacecraft to make the same trip.

- Earth has a diameter of 12,000 kilometers. Calculate how long it would take light to travel around the planet once. Compare that to how long it would take the spacecraft.

Hubble 1

The Hubble telescope is truly our eye above the sky. Light that has traveled millions and even billions of light years without being disturbed is often lost to our ground-based telescopes in the final 80 kilometers of its journey through our atmosphere. The Hubble telescope avoids that problem by orbiting 600 kilometers above the earth and its atmosphere. The radius from the center of the earth to the Hubble telescope is 7,000 kilometers.

If the orbital speed of the Hubble telescope is 28,000 kilometers/hour, calculate how long it takes it to orbit Earth once.

35

Famous Scientists I

Many men and women throughout history have made significant contributions to our understanding of nature and how it operates. Some stand out a little more because their contributions changed the way we look at nature and opened up new avenues of discovery. Early scientists who changed the world include Aristotle, Copernicus, Galileo Galilei, Isaac Newton, and Gregor Mendel.

Solve this anagram to identify one of the famous scientists mentioned above.

action as new

36

Which Way Whiplash?

Newton's first law of motion states that objects at rest tend to stay at rest, while objects in motion tend to stay in motion unless a force acts on them. Devices such as seat belts and air bags are used to protect passengers from the consequences of this law when accidents occur.

Another safety device that has been added to vehicles is the head restraint. When a car is hit from behind, Newton's first law comes into play. Many people say that in such an accident without head restraints, the person's head whips backwards, causing the injury known as whiplash. As an observer watching from outside the car, use Newton's first law of motion to explain what really happens to the person's head and why a head restraint would help.

37

© 2003 J. Weston Walch, Publisher

Forced Into It I

Newton's second law of motion states that if a net force acts on an object, that object will accelerate in the same direction as the net force. Another way of looking at it is to say that if an object is accelerating, there must be a net force acting on it.

Circle each situation described below in which the object has a net force acting on it. For each one you circle, identify the direction in which the net force is acting.

38

1. A car moves to the right while slowing down. _____

2. A marble moves in a circular path inside a paper plate at a constant speed. _____

3. The Moon orbits Earth. _____

4. An air hockey puck moves smoothly across the air hockey table after being struck. _____

5. A rocket is launched upward from the launch pad. _____

© 2003 J. Weston Walch, Publisher

Action-Reaction Pairs

Newton's third law of motion states that as object A exerts a force on object B, object B simultaneously exerts an equal force in the opposite direction on object A. These forces are often referred to as action-reaction pairs.

One way to identify the pairs correctly is to think about how the motion of each object is affected. If the forces are opposite in direction, the effects should be opposite. Try using that approach as you complete the table below.

Action Force	Reaction Force	Action Effect	Reaction Effect
golf club hits golf ball	golf ball hits golf club	golf ball speeds up	golf club slows down
Earth pulls ball down		ball accelerates downward	
Sun pulls inward on Earth		Earth orbits the Sun	
boy's hands push against friend's chest		friend falls backward	

39

© 2003 J. Weston Walch, Publisher

Newton's Influence

Isaac Newton (1642–1727) had a tremendous influence on our understanding of the rules by which nature behaves. He was the first to mathematically describe the universal nature of gravity and the effect of forces on motion, and to realize that white light is composed of separate colors. Along the way, he developed calculus to describe his findings mathematically.

In 1730, the poet Alexander Pope wrote the following epitaph for Isaac Newton:

> Nature and Nature's laws lay hid in night;
> God said, Let Newton be! and all was light.

Write a paragraph explaining why you think Pope gave Newton this tribute.

Phamous Phrases (1)

Some of the laws of physics are familiar only to those who have studied physics, while others can be quoted by those who have never been near a physics classroom.

In a fallen phrase puzzle, the letters in each column have fallen in random order from the corresponding column above them. It's your job to fit each letter back into its correct place in the puzzle. Solve the fallen phrase puzzle below to see which famous law of physics is stated.

	F	O	R							
F	O	R	C	E						
					E	Q	U	A	L	
			O	P	P	O	S	I	T	E
					P	O				
	N		F	N	T	E	E	Y		
A	F	O	A	O	P	E	E	U	A	L
F	O	D	R	E	E	C	H	S	I	T
I	S	R	G	O	R	V	Q	R	R	E

41

© 2003 J. Weston Walch, Publisher

When Push Comes to Shove

Newton's third law of motion states that for every action there is an equal and opposite reaction. This law can be used to describe why objects move forward.

Use Newton's third law of motion to complete these descriptions.

1. A fish's fin pushes water backwards.

   The water pushes _____

2. A car's tires push backward against the road.

   The road pushes _____

3. A swimmer pushes backwards against the water.

   The water pushes _____

4. Your foot pushes backwards against the floor.

   The floor pushes _____

42

Phamous Phrases II

Newton's three laws of motion are a cornerstone of classical physics. In essence, they describe any and all motion by explaining what happens to an object when no forces act on it and when forces do act on it, and what happens to the object exerting the force.

Solve the double puzzle to reveal a statement of one of Newton's three laws of motion.

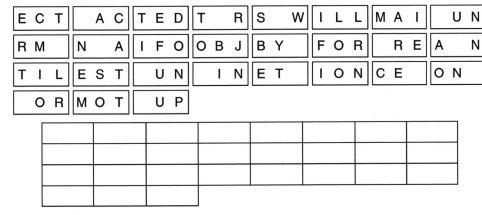

E C T		A C T E D	T	R S	W I L L	M A I		U N
R M	N	A	I F O	O B J	B Y		F O R		R E A	N
T I L	E S T		U N		I N E T		I O N	C E		O N
	O R	M O T		U P							

43

Phamous Phrases III

Isaac Newton is perhaps associated with more well-known breakthroughs in physics than any other person in history. His laws of motion and gravity, along with his discoveries regarding light and color, are still the basis of most introductory physics courses taught today. His personal history shows him to have been a very complex person who did not get along particularly well with others.

In one of his more appreciative moments, Newton made a very sincere remark which has served to point out that scientific discovery is often a combination of many things and many people. Solve the letter tile puzzle to complete Newton's famous quote, which begins

"I have seen further than others by _____ ."

STA	IAN	SH	TS	DER	THE	ON	S	O

NDI	OUL	NG	F	G

44

Which Way to Go?

The forces that act upon any object at any given time determine the motion (or lack of motion) of that object. The key to analyzing that motion is to correctly identify each of the forces acting and the direction in which it is acting.

In each situation described below, draw and label a force vector for each force that is acting on the <u>underlined</u> object. Each object is represented by a box.

1. A <u>box</u> is sitting on a table.

2. A person lifts a <u>laundry basket</u> upward

3. A <u>sled</u> slows down as it slides across the street in this direction:
   →

Cruisin'

The cruise control on a car works to keep the car moving at a constant speed. According to Newton's laws of motion, this means that the net external force acting on the car is zero. However, this does not mean that that there are no forces acting on the car, just that they all cancel each other out.

On the drawing of the car below, draw and label an arrow to represent each force acting on the car as it travels along the highway using its cruise control to achieve constant speed.

46

Mass or Weight?

Two quantities that are often confused are **mass** and **weight.** Perhaps this is because they are so closely tied together. Mass is a measure of an object's inertia, while weight is a measure of the gravitational force exerted on an object by the earth.

The relation between the mass of an object and its weight on Earth is given by the equation $F_w = mg$ where g is the acceleration due to gravity (9.8 m/s^2). Complete the table below.

Mass (kg)	Weight (N)
1	
	1
25	
	25

© 2003 J. Weston Walch, Publisher

Weigh Down Deep!

Newton's universal law of gravitation can be used to calculate the weight of a person on Earth. Since weight is the amount of gravitational force acting between the person and Earth, it can be found by using $F_{weight} = F_{gravity} = G\,(m_{person} \cdot m_{earth})/(r_{earth})^2$ where G is the constant of universal gravitation.

Suppose you could dig a tunnel from Earth's surface straight to the center of Earth. As you crawled down the tunnel, would your weight increase, decrease, or stay the same? Defend your answer.

48

Phamous Phrases IV

History books often include famous quotations from well-known people. These quotations usually define their places in history and reflect their accomplishments. But what if history has failed to record the words of someone famous at the time of his or her great discovery? Couldn't we guess what the person might have said?

Solve the letter tile puzzle to see what Isaac Newton might have said while he studied gravity.

| E | | D | C O M | | M U | | U P | O E S | O W N | T | | G | S T |
| W H A | | | | | | | | | | | | | |

| | | | | | | | |
| | | | | | | | |

49

Moon Droppings

In 1971, Apollo astronaut David Scott performed a simple experiment for students and teachers everywhere. He dropped a hammer and a feather from the same height above the moon's surface at the same time. Upon observing the result, Scott exclaimed, "Mr. Galileo was correct!"

Explain what he observed and why it was different from what he would have observed if he had performed the experiment on Earth.

50

Weight Here!

The weight of an object can be determined from its mass by use of the equation $F_{weight} = mg$ where g is the acceleration due to gravity at that location.

For each of the locations below, calculate the weight of a person whose mass is 50 kg.

Location	g at Location	Weight
1. Earth	9.8 m/s²	_____ newtons
2. Moon	1.6 m/s²	_____ newtons
3. Jupiter	24.9 m/s²	_____ newtons

51

© 2003 J. Weston Walch, Publisher

Lunar Fun

The gravitational field strength at the surface of the Moon is only one-sixth as strong as it is at the surface of Earth.

Suppose a roller coaster, identical to one on Earth, was constructed on the moon inside a small domed city. Describe how riding on the lunar roller coaster would be different from riding the identical coaster on Earth.

52

What's Your Mass and Weight?

Most people know their own weight in pounds, but few know what it is in newtons. Even fewer know their mass in either English or SI units. In fact, because the English unit for mass is rarely used, very few people know that it is the **slug**.

Since mass and weight are related by $F_w = mg$, you can find how many slugs of mass you have by determining your weight in pounds and using the English value for g, which is 32 ft/s². Using the same equation to find your weight in Newtons requires knowing your mass in kilograms and using the metric value of g, which is 9.8 m/s².

Hint: One kilogram of mass weighs 2.2 pounds.

Using your own weight in pounds, calculate both your mass in slugs and your weight in Newtons.

53

Center of Gravity 1

Have you ever tried to balance an object on your finger or with the palm of your hand? If it is something symmetrical, like a basketball, you put your finger at the bottom directly under its center. But if it is an oddly shaped object, like a chair, you have to tilt it. In order to balance something at one point, you have to support it so that its **center of gravity** is directly above or below the point of support. The center of gravity of an object is the point around which all of its mass is considered to be centered.

For each of the objects drawn below, place an **x** at the spot where you think its center of mass is located.

bowling pin   wooden meter stick   golf ball   golf tee   donut

Center of Gravity II

When an object is supported by three or more points, those points define its support base. An object will not fall over as long as its center of gravity remains within its support base. The larger the support base, the more stable the object is. Another factor affecting stability is how high or low the center of gravity is. A lower center of gravity results in greater stability.

Road tests have shown that SUVs have a greater tendency to roll over when traveling on a banked curve than do standard passenger cars. Study the drawings below and then write an explanation of why this is true.

SUV   passenger car

flat road   banked curve

55

Spring Constant

The **spring constant** of a spring is the ratio of the amount of force applied in stretching or compressing a spring to the distance through which it stretches or compresses. It is measured in units of newtons/m.

For each application, judge whether the springs needed should have a relatively high or low spring constant.

	Application	High or Low
1.	the spring in a retractable ink pen	_____
2.	the springs in a car seat	_____
3.	bedsprings in a soft mattress	_____
4.	bedsprings in a hard mattress	_____

56

Directional Spelling

When describing circular motion, three terms are often used to describe direction. The word **centrifugal,** which means outward, is often misused when describing the force acting on the object. The word **centripetal,** which means inward, correctly describes the direction of the force and resulting acceleration. The word **tangent** identifies the direction of the object's velocity at any instant along the circular path.

The diagram below is a neat way of getting the point across that the term centripetal refers to an inward direction.

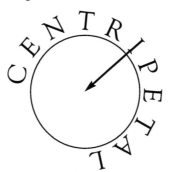

Create diagrams for the terms *centrifugal* and *tangent* that show their directions.

An Inward Look

A **centripetal force** is not a particular type of force, such as gravity, but rather a force that is acting in a particular way. Any force that acts on a moving object in such a way that its direction is always perpendicular to the direction in which the object is moving can be described as a centripetal force. This will cause the object to move in a circular path and the force will always be directed inward toward the center of that circular path.

For each of the situations described, complete the table by identifying the object that is exerting the centripetal force.

Situation	Object Exerting the Centripetal Force
A ball is tied to a string and swung in a horizontal circle.	the string
The clothes and water inside a washing machine are spun in a circle during the spin cycle.	
A car traveling on the road follows a curve in the road.	
A marble is rolled around the inside edge of a paper plate.	
A penny rolls on its edge on the inside of a balloon.	

Where Did the Water Go?

The end result of the spin cycle in a washing machine is that the water has left the tub and the clothes are all pressed against its side. A common, but totally incorrect, explanation is to say that while the water and clothes were in circular motion, an *outward*, centrifugal force was acting. The water was pushed through the holes in the tub wall while the clothes were pushed against the wall. Actually, in all circular motion, the objects traveling in a circular path do so because an *inward*, centripetal force is acting on them.

Explain what force or forces act as a centripetal force (inward) during the spin cycle and how the water and clothes end up where they do. Use a diagram to help in your explanation.

59

© 2003 J. Weston Walch, Publisher

Meltdown

Archimedes' principle states that when an object floats it displaces an amount of fluid whose weight is equal to the weight of the object.

Suppose that a glass containing ice cubes is filled completely to the brim with water such that several of the ice cubes are floating above the rim of the glass. If the ice is allowed to melt completely, will the water level in the glass spill over the rim, fall below the rim, or stay at exactly the same level? Defend your answer.

60

Pool Balls

When an object is floating in a fluid, the fluid is pushing upward on the object with a force called the **buoyant force.** As long as the buoyant force is equal to the weight of the object, the object will float. The amount of the buoyant force is equal to the weight of the fluid that is displaced by the object.

Suppose you are floating in a raft in a swimming pool. On the raft are several bowling balls. If you threw the bowling balls overboard so that they all rested on the bottom of the pool, would the water level along the side of the pool go up, go down, or stay at exactly the same place? Defend your answer.

61

© 2003 J. Weston Walch, Publisher

Coke Float

When objects have a density less than water, they will float. If their density is greater than that of water, they will sink to the bottom.

A 12-oz can of regular cola will slowly sink to the bottom of a fish tank filled with water. A 12-oz can of diet cola will float (although just a small portion of the diet cola will remain above the water). Explain why this is so. *Hint:* Think about what is different between regular cola and diet cola.

Daily Warm-Ups: Physics

62

U.S.S. Nimitz

The **buoyant force** exerted on a floating object is equal to the weight of the fluid displaced by the object. Archimedes reportedly discovered this fact when he was trying to determine if a king's crown was made of pure gold.

The aircraft carrier U.S.S. *Nimitz* weighs over 91,000 tons. If one gallon of seawater weighs 8.328 pounds, calculate the volume of seawater displaced by this gigantic vessel.

63

The Plimsoll Line

The **Plimsoll line** is a set of lines marked on the hull of a ship to show the highest water lines at which the ship may safely float under different conditions.

Suppose a ship is loaded until the water line is at the highest Plimsoll line. It leaves the northeastern coast of the United States in winter sailing to South America and then up the freshwater Amazon River to its destination. Think about the density of water and then explain why this might be a dangerous voyage.

64

Tire Pressure 1

Pressure is defined as the amount of force per unit area. The four tires on an automobile are responsible for supporting the car's total weight. The surface area of a tire that actually touches the road depends on the weight it is supporting and the amount of air pressure in the tire.

A small SUV weighs 4,200 pounds. Assume that each of the four tires supports one-fourth of the car's weight. If each tire had a tire pressure of 28 lb/in^2, calculate the surface area of each tire that is in contact with the road. Can you explain why tires look flatter on a cold morning?

65

© 2003 J. Weston Walch, Publisher

Tire Pressure II

An interesting method of obtaining a reasonable value for the weight of a car uses a tire pressure gauge, four sheets of newspaper, and a can of black spray paint. The car is driven onto the newspapers such that each tire rests in the middle of a sheet. Spray paint is sprayed entirely around each tire, and the air pressure is recorded for each tire on the newspaper containing its "footprint."

Describe a way to determine the weight of the car using the information obtained above.

66

Wind Currents

Wind is air set in motion due to differences in air pressure across the earth. The simplest air circulation pattern in the earth's atmosphere occurs between the equator and the poles. Air rises at one location and sinks at the other.

For each location, use what you know about the effect of heat on air pressure to answer the questions.

	North Pole	Equator
1. Does air rise or sink here?	_____	_____
2. Why does it rise or sink here?	_____	_____
	_____	_____
3. Is the air pressure high or low here?	_____	_____

67

Lift Me Up!

As an object goes deeper into water, the pressure exerted by the water on the object increases. This is why divers have to be careful about how fast they go down or come up. Too great a change in pressure in too short a period of time can result in serious medical consequences for the diver. For every 31 feet an object goes down, the pressure increases by 1 atmosphere (14.7 lbs/in^2).

The most common way of determining the buoyant force exerted on an object is to measure the weight of the water displaced by the object. However, that is not the only way. Use the information given above to calculate the buoyant force on an object submerged in water if the object is 10 feet wide by 10 feet long by 15.5 feet tall.

Daily Warm-Ups: Physics

68

Hoover Dam

The Hoover Dam on the Arizona-Nevada border is the highest concrete dam in the United States. Standing an incredible 726 feet tall, it is one of our nation's biggest tourist attractions.

The top of the Hoover Dam is only as wide as a road's width, yet the bottom is over 600 feet thick. Use a calculation to explain why this needs to be so. *Hint:* Remember that for each 31 feet that you go deeper in water, the pressure increases by one atmosphere (14.7 lbs/in²).

69

Bernoulli's Principle

Bernoulli's principle states that a fluid in motion, such as flowing water or moving air, exerts less pressure than when it is not moving.

Have you ever stepped into the shower, turned on the water, and suddenly become startled when the shower curtain moved inward and touched your leg unexpectedly? Use the concept of Bernoulli's principle to explain why this would happen.

70

Torque Your Way In

An object rotates because a **torque** acts on it. When you exert a force at the end of a wrench in order to rotate it (along with the bolt it is attached to), you may not have thought about the fact that you applied the force perpendicular to the handle. The longer the wrench's handle, the less force you have to apply, because the amount of torque is equal to the product of the force times the distance to the axis about which it is rotating.

Use the relationship of torque to force and distance to explain the following two small mysteries in your kitchen. Why is it easier to open a cabinet door when the doorknob is at the end of the door than when it is in the middle of the door? Why is a big doorknob easier to turn than a small one?

71

Just a Moment

When we try to change the velocity of an object in any way, we have to overcome its inertia. Likewise, when we try to change the rotational motion of any object that is spinning, we have to overcome its rotational inertia, which is called its **moment of inertia.** The measurement of inertia is simply its mass, while the measurement of its moment of inertia involves both its mass and the distribution of that mass around its axis of rotation. In general, the farther away the mass is located from the axis of rotation, the greater its moment of inertia and the more resistance it has to changing its rotational motion.

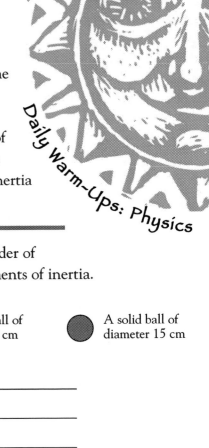

All of the objects below have the same mass. List them in order of easiest to rotate to hardest to rotate based on their moments of inertia.

A solid ball of diameter 30 cm

A hollow ball of diameter 30 cm

A solid ball of diameter 15 cm

(easiest to rotate)   _____

                        _____

(hardest to rotate)   _____

72

Name That Force

Forces are categorized according to the property with which they are associated.

For each of the three forces given below, find the property associated with that force in the word puzzle. *Hint:* Look at the equation used to calculate the force to see which property is included.

Gravitational force	Electrical force	Magnetic force

```
U  J  C  Y  Z  A  M  T  Q  C
U  H  I  E  Q  N  K  V  H  F
L  A  C  N  Q  F  L  A  W  J
A  G  C  D  U  Z  R  U  L  T
T  D  A  O  V  G  J  A  Q  V
E  Q  C  E  E  S  V  Y  O  J
A  A  R  N  L  Q  D  I  L  B
N  S  S  A  M  O  U  Q  G  A
F  T  T  G  P  Q  P  H  Z  M
B  D  P  M  R  I  Z  T  V  J
```

73

© 2003 J. Weston Walch, Publisher

Forced Into It! II

All known forces in the universe seem to fit into one of four categories. We feel the influence of **gravitational** force throughout our entire existence. Most of the other forces we experience are **electromagnetic.** The other two types of forces, **nuclear** and **weak interaction,** are at work inside the nucleus of an atom.

Solve this anagram to identify one of the four categories of forces mentioned above.

to energetic calm

74

The Coanda Effect

When water flows over a curved surface, it has a tendency to stick closely to the surface. This phenomenon is called the **Coanda effect** and has been used in a variety of interesting applications.

A new style of rain gutter has become popular among homeowners who are tired of cleaning out leaves and twigs. The gutters seem to have a solid cover over them, which would certainly keep the leaves out. But how does the rain running off of the roof get into the gutter? Upon closer examination, there is a small gap for the water to run into. Choose which of the designs below is the most efficient at gathering rain water while keeping out leaves and twigs. Explain your choice.

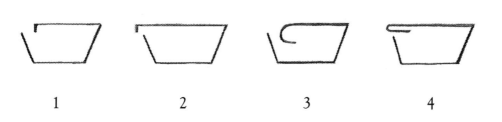

1   2   3   4

75

Work, Work, Work!

The quantity **work** is defined as the product of the force applied to an object multiplied by the distance through which the force is applied. This means that if no displacement of the object occurs, no work is done on the object even though the force applied may be quite large.

Reference to the work we do is a large part of our daily conversation. We all have opinions as to which jobs require more or less work. However, this common usage of the word *work* does not always match up with the physics definition. Use the column on the left to arrange the list of jobs in order of hardest work to easiest in your opinion. Use the column on the right to arrange them in order of most work to least work according to physics.

76

Opinion	Jobs	Physics
(hardest)	store clerk	(most)
	accountant	
	package delivery driver	
(easiest)	furniture mover	(least)

Power Up!

Power is the rate at which energy is used. When an appliance is labeled with a certain power, such as a 1,200-watt hair dryer, it means that during each second of operation the dryer transforms 1,200 joules of energy from one type of energy into other types of energy.

Choose two of the following devices and identify the type of energy that operates the appliance and the type or types of energy it produces: toaster, portable generator, electric dryer, flashlight.

Device	Energy In	Energy Out
_____	_____	_____
_____	_____	_____

77

Stop!

The stopping distance of a vehicle depends upon the amount of work required to bring it to a halt. That amount of work is equal to the change in the vehicle's **kinetic energy.**

Suppose a car and a truck have the same maximum braking force and that the truck weighs twice as much as the car. List each of the five situations below in order of stopping distance required, beginning with the shortest.

Driving Situation	Stopping Distance Rank	
car moving at 30 mph	_____	*(shortest)*
car moving at 60 mph	_____	
car moving at 75 mph	_____	
truck moving at 15 mph	_____	
truck moving at 60 mph	_____	*(longest)*

Lighter Electric Bills

Electric light bulbs come with various power ratings, such as 60-W and 100-W. Since light bulbs are used to produce light, it might seem logical that all light bulbs with the same power rating would produce the same amount of light. However, the power rating refers to how much energy is being *used* each second. Some of that energy is converted to heat as well as light and varies significantly depending on the nature of the bulb.

Incandescent bulbs produce a lot more heat than fluorescent bulbs and therefore use more power to produce the same amount of light. The savings in cost to operate fluorescent bulbs can be quite surprising. It typically costs about 0.007¢ per watt to operate something electrically for one hour. A 60-W incandescent bulb and a 15-W fluorescent bulb give off the same amount of light. Calculate the amount saved over a year's time if they were used an average of 10 hours per day.

79

What's the Matter?

Matter can exist in one of four different states. On Earth, we often deal with only one state of a substance, such as solid aluminum. A few substances, such as water, are used daily in three states. The fourth state, almost nonexistent on Earth, makes up most of our universe.

Find the four states of matter in the word puzzle.

```
S A A P T K D X C S
B L M J O K I L A W
R O P S K L L G K W
G C Z U A V O P Y R
D I U Q I L S C N A
E W W G V G P A E J
C F M W I P L B S Q
B D Y Z B J B L O N
X F N C B K B Z W Y
```

80

To What Degree?

Temperature is often measured using different temperature scales. The **Fahrenheit** scale has long been used in the United States to describe the air temperature in weather reports and for cooking temperatures in recipes. Other countries use the **Celsius** scale for the same applications. Science often has to use another scale, called the **Kelvin** scale, when absolute values of internal energy are to be analyzed. A reading on any of these scales can easily be converted to readings on the other two using the equations $T_F = (9/5\ T_C) + 32$ and $T_K = T_C + 273.15$.

Daily Warm-Ups: Physics

Put each of the temperatures below in order from hottest to coldest.

0 K   0°C   0°F   96°C   100 K   212°F

(hottest)   _____

_____

_____

_____

_____

(coldest)   _____

81

© 2003 J. Weston Walch, Publisher

Loosen Up! I

When heat energy is added to most objects, they expand. For equal changes in temperature, the amount of expansion depends in part on the dimensions of the object as well as the material it is made of. Two rods of different metals but of equal length would expand different amounts due to the difference in the material. Two rods made of the same material but of different lengths would expand different amounts due to the difference in their lengths.

A common way to loosen a metal lid that is screwed tightly to a glass jar is to run hot water over the lid and jar top. Explain why this loosens the lid.

82

Hotter Than You Think

The **Kelvin temperature scale** is sometimes called the absolute scale because a reading of zero truly means the lowest temperature possible. If the temperature of an object were 100 K, then it would need to be raised to 200 K in order for the object to be twice as hot.

On the same scale as originally used, to what temperature would each need to be raised to be twice as hot? *Hint:* $T_K = T_C + 273$

	Original Temperature		Twice as Hot
1.	ice in freezer	–20°C	_____
2.	melting ice	0°C	_____
3.	body temperature	37°C	_____
4.	boiling water	100°C	_____

Cool Feet

Thermal conductivity is the measure of how well different substances conduct heat. When a material with a relatively high thermal conductivity touches an object hotter than itself, heat will flow quickly into the material. As a result, it feels hotter than would a material with a lower thermal conductivity. This is why it is better to roast marshmallows over a fire using a wooden branch than a metal coathanger.

Use the concept of thermal conductivity to explain why porcelain tile feels colder to your bare feet than carpet.

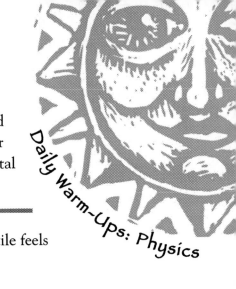

Daily Warm-Ups: Physics

84

The Cooking Glass

Certain materials not only transmit heat well through **conduction,** they transmit it evenly throughout the material. Unfortunately, regular glass is not one of those materials. It does a good job of transmitting heat, it just doesn't do it evenly.

Pyrex glass was specially developed to overcome the problem with uneven heating. As a result, it only expands half as much when heated as regular glass. Explain why Pyrex glass is better than regular glass to use when cooking hot dishes.

85

Hot 'n' Cold I

Heat can be transmitted by three different methods. When hot objects touch cooler objects, heat is transmitted by conduction. As a warmer fluid moves into a cooler region, heat is transmitted by **convection.** The transmission of heat through electromagnetic waves is called **radiation.** Any material that effectively reduces the transmission of heat is called a thermal insulator.

Thermos bottles have been used for many years to keep coffee hot for many hours. They basically consist of double glass walls, which are coated with a silver-colored coating. The air between the double walls has been pumped out, creating a vacuum. Explain the roles that the silver coating and the vacuum play in making the thermos bottle such an effective thermal insulator.

Daily Warm-Ups: Physics

86

Meat Platter

Each decade seems to have its own collection of unique kitchen items sold through television and specialty catalog advertising. One such item sold in the '90s was a "magic" defrosting plate. You simply laid the frozen meat on the plate and in just a few minutes the meat was ready for cooking. Manufacturers claimed to have sold several million of these in just one year.

Several science-oriented publications printed articles describing the defrosting plate phenomenon. Although the various brands differed in size and thickness, they all were similar in that they were simply a metal block painted black. Users were instructed to run hot water over the plate after a few minutes to renew its defrosting abilities.

Identify which method of heat transfer was being used (conduction, convection, or radiation), and explain how running hot water over the plate renewed its ability to defrost.

87

Hot 'n' Cold II

Any material that does not conduct heat well is a **thermal insulator.**
Air is a poor conductor of heat and is the main ingredient in many
thermal insulation materials.

Explain why Styrofoam cups work equally well at keeping hot drinks
hot and cold drinks cold, while still making it possible for you to hold
onto the cup.

88

Sky High Cooking

The boiling temperature for water is dependent on the atmospheric pressure. At standard atmospheric pressure, the boiling temperature is 100°C. At altitudes below sea level, where the atmospheric pressure is greater, the boiling temperature is higher. Altitudes above sea level would result in water boiling below 100°C.

Pressure cookers allow the cook to regulate the air pressure inside the cooker at levels greater than standard atmospheric pressure. When a recipe calls for cooking something in boiling water, it is assumed that the water is at a temperature of 100°C. Suppose you had a cooking thermometer and a pressure cooker available. Explain how you could stay true to the intent of a recipe, which called for cooking in boiling water, if you were in Death Valley, California (significantly below sea level) and Denver, Colorado (significantly above sea level).

89

Up on the Roof

The color of your roof can play a major role in how hot or cold your house becomes when the sun shines on it. Dark colors absorb much of the sun's heat energy, while lighter colors will reflect a good portion of that energy. On hot summer days it would be better to have a light roof, while on cold winter days it would be better to have a darker roof.

Obviously, you can only have one color of roof on your house. For your geographic location, explain which color roof would be better in terms of energy efficiency over the course of a year.

90

Loosen Up! II

Impulse is found by multiplying the force acting on an object by the amount of time the force acts. The amount of impulse that an object experiences is equal to the amount that its momentum changes. This means that the same change in momentum can be accomplished by exerting a large force for a short amount of time or a small force for a longer amount of time.

Seat belts are designed to give a little when they are properly fastened. If not, they would not work as well at keeping you from injury in an accident or sudden stop. Use the concept of impulse equaling the change in momentum to explain why they need to give.

91

Shockingly Simple

Automobiles have many parts that use the concept of **impulse** to reduce the amount of force acting in that area of the car by lengthening the amount of time needed for the force to act in order to accomplish its task. Seat belts and shock absorbers are two such examples.

Identify at least three other parts of an automobile that use the concept of impulse in this way.

92

Body Air Bags

Motorcycle accidents resulting in a fall often produce serious injuries due to the high speeds involved. In Europe, where more people ride motorcycles on roads containing more curves, the accident rate is even higher than it is in the United States. A new device has been produced for sale in Europe that might reduce the number of serious motorcycle injuries. It looks like a fisherman's life jacket and is worn by the rider. The jacket uses sensors to monitor stability. When a fall is foreseen, it inflates in less than 50 milliseconds, creating a body air bag.

Explain the physics principle used by the air bag to protect the rider as he hits the ground. Do you think that this device would protect the rider as well as an air bag in a car? Defend your answer.

93

Are You a Conservationist?

Conservation laws are the cornerstones of physics. When we say a quantity is conserved, we mean that the sum total of that quantity in a system never changes. Quantities that are known to be conserved include momentum, energy, angular momentum, and electric charge. Each has its own conservation law.

Solve this anagram to identify one of the conservation laws in physics.

an uncommon, softer motive

94

Spin Out

One of the quantities known to be conserved is **angular momentum.** The angular momentum of a spinning object will not change unless an outside torque acts on it. The angular momentum of any object can be calculated by multiplying its moment of inertia by its rotational speed.

When an ice skater goes into a spin, he draws his arms inward to decrease his moment of inertia. Since his angular momentum is conserved, his rotational speed increases. Use the conservation of angular momentum to explain why a cheerleader or an Olympic diver goes into a tuck position when beginning a somersault, and comes out of the tuck near the end of the somersault.

The Thrill of Conservation!

Roller coasters are an excellent example of the **conservation of energy.** Work is done to raise the cars to the top of the first hill and then gravity transforms the energy back and forth between potential and kinetic for the rest of the ride.

Cedar Point Amusement Park in Sandusky, Ohio is the self-proclaimed roller coaster capital of the world. One of its 14 coasters is the Magnum XL-200. It stands out above the horizon with an incredible drop of 59.3 meters on the first hill! Disregarding any loss of energy to friction, calculate the speed of the cars at the bottom of that first hill.

96

Collision Conservation

Momentum is always conserved in any collision, as is energy. However, energy is usually transformed into several different types as a result of the collision. In most collisions, a significant amount of the kinetic energy is converted into other forms, and the colliding objects' movements are greatly affected. These are called **inelastic** collisions. There are some collisions, like the colliding steel balls suspended by strings in a Newton's Cradle, in which kinetic energy is almost conserved. Since so little kinetic energy is lost in each collision, they seem to just go on bouncing. These collisions represent **elastic** collisions.

Elastic collisions always seem to catch our attention. Because they are so rare, they seem unusual to us. Use a personal experience to describe an elastic collision that you have seen.

97

© 2003 J. Weston Walch, Publisher

Perpetual Motion I

Perpetual motion machines are machines that, once working, would keep on going by themselves. There are actually two types of these theoretical perpetual motion machines. A perpetual motion machine of the first kind is one in which the machine produces more energy than it uses while running.

Use the laws of physics to explain why a perpetual motion machine of the first kind is theoretically impossible to build.

98

Perpetual Motion II

Although the law of conservation of energy and the first law of thermodynamics do not allow for the possibility of a perpetual motion machine of the first kind, they do not forbid the possibility of a perpetual motion machine of the second kind. This machine would simply create an amount of energy equal to that required to run it, thus opening the possibility of its running itself forever, once started.

Although the laws of physics referred to above do not prohibit such a machine, another law of physics does (the second law of thermodynamics). Use what you know about friction to explain why a perpetual motion machine of the second kind could not work.

99

© 2003 J. Weston Walch, Publisher

Just Charge It

Most objects are electrically neutral. They contain equal amounts of positive and negative charges. However, many objects often become charged as they gain or lose electrons. One way this may happen is through **charging by friction,** in which two neutral objects rub against each other. The resulting transfer of electrons leaves one object negatively charged and the other positively charged.

Describe a common event in which objects become charged by friction.

100

It's a Wrap

Clear food wraps cling to whatever surface they are pressed against (even themselves). When stretched tightly over a container and pressed against the sides, they keep their tension and secure the container. The secret behind these plastic wraps is static electricity. Some of the static charge is created when the roll is manufactured. Since plastic is not a good conductor, the charge remains for a long time. More static charge is created when the plastic is pulled off the roll. Pull faster and it will cling even better.

You might have noticed that these clear plastic wraps do not work as well in humid conditions and hardly work at all when the container has moisture on it. Explain why this is so.

101

Electric Force

An **electric force** is the force that exists between any two charged particles due to their charge. Objects with opposite charges will attract each other, while objects with the same charge will repel each other.

Complete the table for each pair of objects listed.

Pair of Objects	Electric Force Present?	Repel or Attract?
a proton and a proton		
a neutron and a proton		
an electron and a neutron		
a neutron and a neutron		
an electron and an electron		
a proton and an electron		

102

Electric Field

An **electric field** is the region in space around a charged object in which another charged object would experience an electrical force due to the charges. The direction of the electric field at any point is the direction in which a positively charged particle would move if placed at that point. If there is more than one source charge, the electric field at any point is the vector sum of the sources at that point.

Each of the four electric fields shown below are generated by the source charge or charges indicated by the circles. The lines show the electric field lines in that region, and the arrows show the direction of the electric field along those lines. Study the electric field and then indicate whether each source charge is positive or negative by putting a + or a − inside the circle.

103

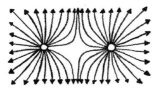

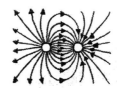

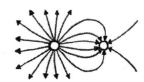

Plugged In

Electrical appliances can be found in practically every room of most homes in the modern world. They help us store and prepare food, provide entertainment, and allow us to live in a comfortable environment. What all of these appliances have in common is that they convert electrical energy into some other form of energy.

Fill in the blanks to identify four of the types of energy produced by electrical appliances.

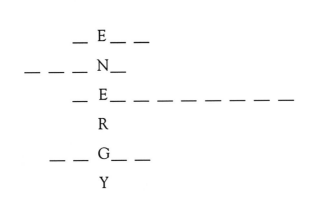

```
_  E __ __
__ __ __ N__
_  E__ __ __ __ __ __ __ __
   R
_  __ G__ __
   Y
```

104

Bright and Brighter

A **series circuit** contains two or more resistance sources in one single path. A **parallel circuit** contains two or more resistance sources, each with its own separate path.

Two standard household light bulbs, one 60-W and one 75-W, are hooked together in a series circuit, and the brightness of each is observed. The two bulbs are then connected using a parallel circuit and once again observed. For each circuit, identify which bulb burns the brightest and explain why.

105

Hush Puppy Overload

Rooms are wired using **parallel circuits** so that multiple outlets can be available when several electrical devices need to be running at the same time. With added convenience comes added danger. As each additional device is turned on, the total current in the circuit increases, as does the risk of overheating the wires. Fuses and circuit breakers are used to limit the amount of current to a safe level.

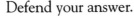

The school's football booster club has decided to have a fish fry as a fundraising event. The fish will be cooked in propane-fueled fryers, while the hush puppies are to be fried in electric cookers. The president of the club has checked the picnic shelter and counted a total of eight electrical outlets that can be used. The circuit provides 120 volts and contains a 40-amp circuit breaker. If each electric fryer has a power rating of 900-W, how many of the outlets can be used before the circuit overloads? Defend your answer.

106

Wired at Home

Every home contains a circuit-breaker box or a fuse box containing several breaker switches or fuses. In fact, each switch or fuse represents a separate circuit in the home. While a few of those circuits contain only one plug for a large appliance, most contain several plugs, which are wired in parallel.

Write a paragraph explaining why wall plugs in your house have to be wired in parallel. Include in your explanation the reason there must be several parallel circuits instead of just one big one.

107

© 2003 J. Weston Walch, Publisher

Magnetic Domains

A **magnetic domain** is a microscopic cluster of atoms that have their fields aligned. Some metals, such as iron, contain magnetic domains. They most often are not magnetic because their domains are randomly aligned. However, when the domains are aligned, the iron becomes a magnet.

A metal rod and a tiny rock,
   Are sitting on the ground.
A man then takes the rod and rock,
   And begins to pound.
A long time later he takes the rod,
   And sticks it near another.
The metal rod, points down to nod,
   Attracted to its brother.

—*Jason Vincent*

108

Explain the physics behind this description of the making of a magnet.

Magnetic Force

A **magnetic force** is the force created by two magnetic fields. Fields with opposite directions create an attractive force, while fields in the same direction create an opposing force.

Put an **x** in the space next to each item below in which a magnetic force exists. For those items in which a magnetic force is not present, explain why not in the space next to it.

Bar Magnet

| N   S |

1. _____

Wire

I = 0

v = 0   (e⁻)   electron

2. _____

Horseshoe Magnet

Wire

I = 3 A

3. _____

v = 2 × 10⁶ m/s   (e⁻)   electron

4. _____

Wire

I = 2 A

Wire

I = 2 A

109

Magnetic Field

A **magnetic field** is created whenever a charged particle moves. A permanent magnet has a magnetic field because the movement of electrons in the majority of the atoms is aligned in the same direction.

Put an **x** in the space next to each item below in which a magnetic field exists. For those items in which a magnetic field is not present, write the reason why in the space next to it.

Bar Magnet

| N | S |

1. _____

Wire

I = 0

2. _____

110

Wire

I = 2 A

3. _____

Horseshoe Magnet

N   S

4. _____

Magnet Earth 1

An **electromagnet** differs from a bar magnet in its origin. A bar magnet's field strength is steady due to the stationary nature of its magnetic domains. The magnetic field of an electromagnet can be varied simply by changing the electric current running through its wires. Vary the amount of current, and the strength of the field changes. Reverse the direction of the current, and the direction of the magnetic field reverses as well.

Earth's magnetic field looks like that of a bar magnet. However, its origin is believed to be the movement of liquid iron and metal moving in Earth's core.

Do you think Earth's magnetic field is more like that of a bar magnet or that of an electromagnet? Explain your reasoning.

111

Magnet Earth II

Earth's magnetic field surrounds the planet, and its magnetic poles are located near our geographic poles. Scientists believe that Earth has had its magnetic field for at least 3 billion years and that the poles have reversed themselves many times over that time period. Earth's magnetic field is important in shielding us from cosmic rays. The field is thought to be the result of the movement of liquid iron and metal in Earth's core.

Daily Warm-Ups: Physics

If some geological catastrophe caused a change in the way liquid metal moved in Earth's core, a noticeable change in Earth's magnetic field would occur. In a modern society that depends heavily on electromagnetic radiation for a wide variety of daily functions, many of those functions would be affected. Describe two such disruptions that might occur.

112

Maxwell's Equations

In the 1870s, James Clerk Maxwell summarized all that was known about electricity and magnetism with four basic equations. Bypassing the mathematical formulas, they can be stated as follows.

1. Unlike charges attract each other; like charges repel.
2. Magnetic poles cannot be isolated from each other.
3. Electric currents create magnetic fields.
4. Changing magnetic fields can create electric current.

For each of the four statements of Maxwell's equations, complete the table by giving an example or occurrence of each.

Maxwell's Equation	Example or Occurrence
1. Unlike charges attract each other; like charges repel.	
2. Magnetic poles cannot be isolated from each other.	
3. Electric currents create magnetic fields.	
4. Changing magnetic fields can create electric current.	

113

© 2003 J. Weston Walch, Publisher

Good Vibrations

The source of all waves is a vibrating object.

Complete the table below by identifying the source of each wave described.

Wave	Source
A tuning fork is struck with a rubber hammer, producing a sound wave.	
A motorboat moves through the water, leaving its wake behind.	
A performer sings a high note.	
A light bulb gives off light.	

114

The Wave

Waves can be classified in two different ways. One way is by using the nature of the wave's disturbance. The other way is by using the direction in which the disturbance is taking place.

Find and circle the names of the two types of waves in each of the two types of classifications. *(There are four names in all.)* You may find the answers by reading forward or backward.

wavelectromagneticsoundlanidutignolightmechanicalightransversenergy

© 2003 J. Weston Walch, Publisher

Type Casting

Waves that cause a medium to be disturbed in a direction perpendicular to the direction in which the wave is traveling are called **transverse** waves. When the medium is disturbed in a direction parallel to the direction in which the wave is traveling, the wave is called **longitudinal.**

Complete the table below by identifying each wave as being either transverse or longitudinal.

Wave	Type
a sound wave	
a water wave caused by a boat moving	
a wave in a rope caused by one end being moved up and down	
a wave in a coiled spring caused by pushing one end in and out repeatedly	
a light wave	

116

Foot Stompin'

Resonance occurs when the frequency of a forced vibration on an object matches the object's natural frequency. This causes a great increase in amplitude, which increases the power transmitted by the object. In 1940, the Tacoma Narrows suspension bridge collapsed when wind-driven oscillations produced resonance in the bridge. Films of its collapse have become favorites among physics teachers and their students. Subsequent designs have incorporated such innovations as separate parallel roadways as a way to keep this type of disaster from happening again.

In the 1800s, English soldiers marching across a small suspension bridge caused it to collapse when their marching set it into resonance. Their marching was in rhythm with the bridge's natural frequency. Since that time, soldiers and marching bands have been told to not march in step when crossing any type of suspension bridge.

Give another example of the disastrous effects of resonance and describe how it happens.

117

Cheap Seats

Modern baseball stadiums have a lot more to offer fans than the stadiums of decades past. One thing they still have in common, however, are the "cheap seats." The back row of the upper deck will always be a long way from home plate. Not only is it hard to see what's going on, it is even hard to hear what's going on.

Suppose you are sitting in your cheap seat and you decide to listen to the broadcast of the game on your pocket radio. When the first batter makes contact with the ball, you hear the crack of the bat through your radio earphones before you hear it in the ballpark! Explain why.

118

Did You Hear What I Saw?

Lightning and thunder are created at the same time by the same event. The speed of light is 186,000 miles per second, while the speed of sound in air is typically around 700 miles per hour. As a result, you see the lightning before you hear the thunder.

You might have heard the old rule of thumb that every five seconds that elapse between seeing the lightning and hearing the thunder represents one mile of distance between you and the lightning. Use the speeds of light and sound to check the accuracy of that statement.

119

© 2003 J. Weston Walch, Publisher

The Didgeridoo

The **didgeridoo** is a musical instrument used by the Aborigines of Australia. It is made of wood that has been hollowed out to create a tube that is open on both ends. When vibrations are made with one's lips near one end, a standing sound wave is created in the tube. This wave produces a haunting, low-frequency note. Different-length tubes will produce notes of different frequencies. The richness of the sound is supposedly enhanced by using a tube that was hollowed out by termites.

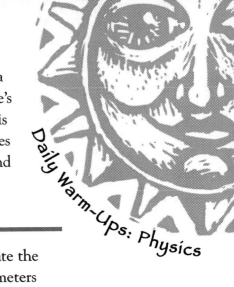

Daily Warm-Ups: Physics

Using 340 meters per second as the speed of sound in air, calculate the frequency of the sound produced in a didgeridoo that is 1.2 meters long. *Hint:* Remember that for a standing wave to occur, there must be an antinode (largest amplitude) at an open end.

120

Can You Hear Me Now?

The loudness of sound can be measured in units called **decibels (dB).** The sound level of a person talking in a normal conversation is typically in the 60-dB range. Continued exposure to sound levels over 85 dB can cause permanent ear damage. People who work in loud environments, such as factories with large machines, are required to wear ear plugs.

Modern technology has made it possible to play music in car stereos that far exceed 85 dB. Many teenagers spend significant amounts of money to raise the decibel level of their car stereos to 120 dB and beyond. As a result, some communities have passed laws limiting the sound level at which music can be played. Write a paragraph expressing your view on the need for such laws.

121

Speedy Clue

Objects can be identified in a variety of ways. We can use either an object's physical properties or its behavior as a clue to what it is.

Identify the unknown item described in the verse below.

> Always at the same velocity
> Ten to the eighth it is, times three
> Faster than sound, and faster than thought
> Nothing is faster, and nothing's a lot
>
> *—Jason Vincent*

122

Electromagnetic Spectrum

The **electromagnetic spectrum** is divided into different groups according to the range of frequencies and wavelengths of the electromagnetic waves. What we commonly refer to as light waves comprise just one of these groups.

Solve the word puzzle to find the five main groups of electromagnetic waves.

```
H M Q U L H N N U N H D
G K I I D I S B I N A U
L I G C A M M A G H Z A
N H J Z R L C H M H S F
T Z S H A O L B Y I P Y
C C Y P D X W E H U W A
B R D J I N L A P U Y R
Y B O K O V T V V T I X
P A T O I O U G L E S O
```

123

© 2003 J. Weston Walch, Publisher

Radar

Radar, invented in the 1930s, sends and receives electromagnetic waves in the radio and microwave wavelengths. It has become very sophisticated over the past two decades, enabling such things as accurate speed measurements of tennis serves and down-to-the-minute predictions of a tornado's path.

The word *radar* is an acronym made from the first letters of a phrase that describes what it does. Solve the letter tile puzzle below to discover the phrase from which radar got its name.

124

| D E T | E C T | D | R I O | I N G | R A D | A N | A N G |
| I O N |

The Long and Short of It

Radio waves are **electromagnetic waves** and, therefore, travel at the speed of light (3×10^8 m/s). When you are driving in your car, AM radio reception fades out when you go under an overpass, while FM radio reception does not.

Local AM radio station WKCT is found at 930 kHz, while FM station WUHU is located at 107.1 MHz. Calculate the wavelength of waves transmitted by each station. Use the results of your calculations to develop an explanation of why AM stations fade out under overpasses, while FM stations do not.

125

Time Will Tell

The atomic clock at the National Institute of Standards and Technology in Colorado sends out signals every second, giving the official time. Clocks and wristwatches are now being made that receive the signal and are, therefore, always accurate to the exact second!

Consider someone on the East Coast who buys one of these watches. He or she would be about 3,200 kilometers from Colorado. Will the fact that the radio signals (moving at 3×10^8 kilometers per second) have to travel that distance affect the one-second accuracy of the watch? Defend your answer with a calculation of the time delay between Colorado and the person wearing the watch.

126

The Speed of Light Quandary

Electromagnetic waves are a disturbance in the electromagnetic field and travel at a constant speed of 3×10^8 miles per second. Simply put, the speed of light is a constant throughout the universe. However, in the study of refraction, we use the fact that light "slows down" as it enters a transparent material to explain why it refracts. We even use its index of refraction to calculate the speed of light through that material.

There seems to be a major quandary here. How can we say that the speed of light is invariant, and yet talk about its slowing down as it goes through a transparent substance? The answer lies in the make-up of transparent materials. The atoms absorb and then re-emit the photons of light as they go through the material. Use this information to explain what we mean when we say that light slows down.

127

The Human Detector

Electromagnetic radiation is usually grouped into parts according to its wavelength, frequency, and energy. One of those groups, visible light, can be detected by the human eye.

The human body can detect other kinds of electromagnetic radiation. Give an example of two types of electromagnetic radiation, other than visible light, that can be detected.

128

Laser

Lasers produce coherent light in an intense, narrow beam. Because of their properties, lasers are ideal for a wide variety of applications. They are used by doctors performing precision surgery, by surveyors measuring distances, and by DVD and CD players.

Solve the fallen phrase puzzle below to discover the phrase from which the laser got its name.

A	M	P	L	I	F	I	C	A	T	I	O	N
									O	F		
				B	I		A	A		O		
		M	L	L	F	L	A	T		I		
	S	T	A	U	I	Y	H	I	O	D	F	
	M	R	I	S	A	O	N	T	D	O		
A	E	P	I	M	I	G	T	T	E	N	O	N

129

Color

The color of an object is determined both by the colors of light that illuminate the object and the colors of light that are absorbed by the object. The colors that aren't absorbed are the ones that reflect back into our eye and therefore determine what we perceive. A red shirt appears red when illuminated with white light because the blue and green portions of the spectrum are absorbed, while red is reflected. If that same shirt were illuminated with just blue light, it would appear black.

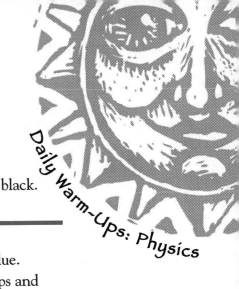

Group the visible spectrum into three parts—red, green, and blue. Chlorophyll absorbs photons from two of those three groups and converts their energy into food for the plant. Identify which two groups are absorbed and explain why.

130

Roy G. Biv

ROY G. BIV is a colorful guy! Of course, Roy doesn't really exist. Rather, he is our acronym for the seven basic groups of colors of light as arranged by wavelength and frequency.

Solve the word puzzle to find the seven basic color groups of light.

```
R  B  Y  E  X  O  N  Z  R  W
E  K  G  D  G  E  G  Z  N  O
D  U  N  I  F  N  X  K  N  L
W  J  D  E  X  S  A  Y  J  L
A  N  R  N  M  J  X  R  F  E
I  G  R  E  E  N  Q  P  O  Y
J  P  H  L  R  B  P  B  D  X
I  W  X  B  I  J  L  Z  O  C
N  K  X  Q  O  O  T  U  M  D
R  T  T  E  L  O  I  V  E  D
```

131

© 2003 J. Weston Walch, Publisher

From One Extreme to the Other

ROY G. BIV helps us to remember the order in which colors of light are arranged in the visible spectrum. It also give us the order of colors as arranged by wavelength, frequency, and energy.

Invisible wavelengths of light that come just before **R** (red) are called *infrared* light, while those just past **V** (violet) are known as *ultraviolet*. The prefixes *infra* and *ultra* refer to how those segments of the electromagnetic spectrum relate to visible light. Think about which aspect of the order of colors (wavelength, frequency, or energy) the prefixes relate to, and then write an explanation of why their use is appropriate.

132

Colors of Light

The three **primary colors of light** are red, green, and blue. When all three are added together, white light is produced. Yellow light is obtained when only red and green are put together; adding red to blue produces magenta; and cyan is the result of putting together blue and green.

When two colors are added together to produce white light, they are called **complementary colors.** Use the information given above to determine the complementary color for each of the colors given below.

	Color	Complementary Color
1.	magenta	_____
2.	cyan	_____
3.	yellow	_____

133

Primary Colors

White light can be produced by mixing proper amounts of just three colors of light. These three colors are called the **additive primary colors.**

Solve the word puzzle to find the three additive primary colors of light.

```
N Y X E E F S Q J R
K E O W O O X R T Q
A M E H N D Z J E S
P V K R Y Z T O U Z
S C D U G V D K L S
U V M E L E F Q B B
B U Y H R Y M E C Q
K N Y P C Z F Y Z V
```

134

Stained Glass

The color of a transparent material, such as stained glass, is determined by the colors of light that are transmitted by the glass. When white light strikes a red piece of stained glass, the blue and green parts of the spectrum are absorbed, while the red part travels through and into your eye.

During the morning events, the stained glass windows in a church look bright when viewed from either the inside or the outside of the church. However, at night they look gray from the inside but still bright from the outside. Explain why.

135

Acting Lightly

When light strikes the boundary of a new medium, one or more of four things can happen. The light can bounce off the medium (**reflection**), be absorbed into the medium (**absorption**), change direction as it goes through the medium (**refraction**), or bend around the edge of the medium (**diffraction**).

Solve these two anagrams to identify two of the types of behavior mentioned above.

spot a robin

nice or left

Lighten Up

Reflection results when light bounces off an object. **Refraction** results when light passes through something transparent.

Look around the room and identify three objects on which light is reflecting and three objects through which light is refracting.
Hint: Some objects might be an example for both!

137

Behaving Lightly

When a ray of light reaches the boundary of a new medium, one of four things can happen to it. In one case, the light energy is converted into heat energy. Second, the light may bounce back and continue moving in the original medium. Third, it may move into and through the new medium at a different speed. Fourth, the light path may bend as it strikes the edge of a new medium.

Solve the word puzzle to find the names of these four behaviors of light.

138

```
F  N  D  C  K  Z  W  T  J  O  N  R
A  Y  T  Q  K  L  Z  G  W  O  D  E
S  S  T  B  Q  J  Y  D  I  G  V  F
R  T  H  K  E  Z  Z  T  Q  U  L  L
W  H  J  M  Q  U  C  F  O  R  X  E
N  O  I  T  C  A  R  F  E  R  C  C
A  B  S  O  R  P  T  I  O  N  R  T
I  J  U  F  P  R  E  C  P  K  G  I
F  N  F  B  D  G  O  O  D  D  S  O
C  D  I  F  F  R  A  C  T  I  O  N
```

Image Is Everything

Real images are formed by converging light rays and can be formed on a screen. They always appear upside down with respect to the actual object.

Virtual images are formed by light that does not actually come to a focus. They can be seen by an observer but cannot be focused on a screen. They always appear right side up with respect to the actual object.

Identify which type of image is being formed by each of the following objects:

Image formation	Type of Image
1. Your bathroom mirror	_____
2. The rearview mirror in your car	_____
3. The side-view mirror on a truck	_____
4. The lens in a camera	_____

139

Screen Saver

Recall that real images are always seen upside down with respect to the object, while virtual images are always seen as being upright with respect to the object.

When you go to a movie, the projector's lens system projects images of the film onto the screen. Identify whether you are watching a real or virtual image. Explain your answer.

Daily Warm-Ups: Physics

What You See Is What You Get

Real images are formed by converging light rays, which can be displayed on a screen or a section of undeveloped film. They always appear upside down with respect to the object. Virtual images, which are formed by using the refraction or reflection of light, cannot be projected onto a surface because light from the object is actually diverging. Virtual images always appear right side up with respect to the object.

The lens of a camera must form a real image onto the film in order to expose it. This means the image formed is upside down. Yet, when you look through a camera, you see the image right side up. Write a possible explanation for how a camera works.

141

Standing Tall

The **law of reflection** is one of the simplest laws in all of physics. It states that the angle at which a light ray is reflected from a surface is equal to the angle at which it struck the surface. Both angles are measured to the normal of the surface.

Look into a plane mirror. In order to see a full-length view of yourself, how tall would be the mirror have to be with respect to your height? Would it be a quarter of your height, half your height, exactly your height, or twice your height? Use a drawing to explain your choice.

Daily Warm-Ups: Physics

Reflect on This

The image in a plane mirror is always virtual, the same size as the object, and reversed (as in left to right). This is why the word "AMBULANCE" is printed backwards on the front of those vehicles. If you were to look into your rearview mirror, you would see its mirror image and know to move out of its way.

Palindromes include words that are spelled the same backwards and forwards. If you printed a palindrome on a piece of paper and held it in front of a plane mirror it may or may not look the same. This is because the letters themselves would also have to look the same when reversed. The words *wow* and *mom* look the same as their image when held in front of a plane mirror. List as many words as you can that would look the same as their mirror image.

143

Plane and Simple

All motor vehicles are required to have a rearview mirror of some type. The typical mirror inside a car or truck is a plane mirror, while many side-view mirrors are actually convex mirrors. Concave mirrors are never used as rearview or side-view mirrors on cars or trucks.

Suppose you had a plane mirror as your rearview mirror and a convex mirror as your side-view mirror. Which one would be the best one to use when you want to determine if it is safe to change lanes? Explain your answer.

144

A Drop in the Bucket

Refraction is the change in direction made by a light ray as it enters into a new medium. When light enters a medium that is optically more dense, it refracts toward the normal (the line it would take if it continued at the same angle). If it enters a medium that is optically less dense, it refracts away from the normal.

You may have been to a restaurant recently where a familiar game sat on the countertop, near the cash register. You drop a coin through a slot and try to hit a small container under water. Winners usually get a free menu item, and the game's profits go to charity. The reason it raises money for the charity is that there are so few winners. Suppose you are viewing the container from your position in front of the counter. Should you try to force the coin to go directly downward from over the container, go in front of the container, or go behind the container? Defend your answer.

Daily Warm-Ups: Physics

145

Can You See Me Now?

Eyeglasses use lenses to correct problems experienced by people who have trouble seeing clearly. Converging lenses are used to correct for people who are farsighted, while diverging lenses are used to correct for those who are nearsighted.

For a lens to converge light, it must have a convex shape. Conversely, for a lens to diverge light, its shape must be concave. However, all eyeglass lenses have a convex shape on the front side and a concave shape on the back. Explain how one pair of eyeglass lenses can end up being converging while another can be diverging.

146

The Eyes Have It

The **focal length** of a lens is determined by both its shape and the material from which it is made. The more curved a lens is, the shorter its focal length. The lens in the human eye can be made to change its focal length. A ring of muscles around the eye is used to contract and thicken the lens. When you look at something close up, the muscles contract and the lens becomes more curved. When you look at something far away, the muscles relax and the lens becomes thinner and less curved.

Accommodation is the ability of the eye to adjust its focal length to see things clearly at different distances. Explain why babies can easily look at something, such as a rattle, that they hold right in front of their eyes, while their grandparents have to hold the newspaper at arm's length to read it.

147

The Wet Look

Eyeglass lenses are made to work when surrounded by air. Converging lenses, which bring light rays together more quickly, help people who are farsighted see more clearly. Diverging lenses, which spread light rays apart, are used by people who are nearsighted.

Suppose a person who normally wears glasses to correct one of the above conditions notices that without glasses he sees more clearly under water than he does in air. Is this person nearsighted or farsighted? Explain your reasoning.

148

Phamous Phrases V

Albert Einstein is most famous for his theory of relativity. The ideas expressed in that work have completely changed the way in which we view the universe. However, several of his quotations about the way we view learning have become an inspiration to many who are turned off by the memorization of facts.

Solve the cryptogram to reveal one of Einstein's famous quotations.

A	B	C	D	E	F	G	H	I	J	K	L	M	N	O	P	Q	R	S	T	U	V	W	X	Y	Z
								23				17	21	3											

$$\underset{23}{I}\ \underset{17}{M}\ \underset{1}{}\ \underset{9}{}\ \ \underset{23}{I}\ \underset{21}{N}\ \underset{1}{}\ \underset{5}{}\ \ \underset{23}{I}\ \underset{3}{O}\ \underset{21}{N}\ \ \underset{23}{I}\ \underset{20}{}\ \ \underset{17}{M}\ \underset{3}{O}\ \underset{18}{}\ \underset{4}{}$$

$$\underset{23}{I}\ \underset{17}{M}\ \underset{6}{}\ \underset{3}{O}\ \underset{18}{}\ \underset{5}{}\ \ \underset{1}{}\ \underset{21}{N}\ \underset{5}{}\ \ \underset{5}{}\ \underset{11}{}\ \underset{1}{}\ \underset{21}{N}\ \ \underset{19}{}\ \underset{21}{N}\ \underset{3}{O}\ \underset{7}{}\ \underset{14}{}\ \underset{4}{}\ \underset{16}{}\ \underset{9}{}\ \underset{4}{}$$

149

Phamous Phrases VI

Einstein's **special theory of relativity** is based upon two fundamental assumptions, which are called postulates. In order for these postulates to be true, Einstein realized that we had to change the way in which we viewed the universe.

Solve the letter tile puzzle to state one of the fundamental postulates of the special theory of relativity.

GHT		IS	ANT	ERV		SP		LI	EED	THE
LL		NST	OF		CO	OBS	ERS	R	A	FO

Phamous Phrases VII

Albert Einstein was once asked to define relativity for the layperson. After thinking for a few moments, this was his reply:

"When you are courting a nice girl, an hour seems like a second. When you sit on a red hot cinder, a second seems like an hour. That's relativity."

Describe in your own words what Einstein's quotation means to you in terms of the measurement of time.

151

Phamous Phrases VIII

Perhaps the most recognizable equation in physics comes from Einstein's theory of relativity: $E = mc^2$. It is the cornerstone of understanding nuclear energy reactions and has guided astrophysicists in their development of the Big Bang theory.

Einstein even liked to answer questions about life analytically. Once, when asked for advice about how to be successful, he replied "If **A** is success in life, then **A** = $x + y + z$. **Work** is x, y is *play,* and z is

_ _ _ _ _ _ _   _ _ _ _   _ _ _ _ _   _ _ _ _."

Solve the letter tile puzzle to find out what z is.

152

| G | | Y | U T H | | S H | O U R | K E E | U T | | M O | P I N |

| | | | | | | | |

Famous Scientists II

Men and women are still making discoveries that totally change our ideas about certain areas of science, revise our theories, and in some cases, abandon centuries-old explanations. Most of those making the discoveries had no idea where technology would take their newfound knowledge. Such was the case with Ernest Rutherford, Neils Bohr, and Enrico Fermi and their contributions toward our understanding of the atom.

Solve this anagram to identify a famous scientist not mentioned above.

sent elite brain

153

Einstein to Bohr

Albert Einstein and Niels Bohr had a famous argument concerning the basic nature of radioactivity. Scientists knew the half-life of several radioactive elements and could therefore determine the time after which one-half of a sample's atoms would decay. They had no idea, however, which particular atoms in the sample would be the ones to decay. Einstein said that there must be an unknown physical mechanism that determined which atom decayed next, while Bohr argued that the whole process was purely random with no mechanism needed. Their public argument, carried out through published letters, ended with a famous quotation from each.

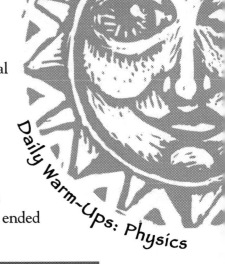

154

Solve the letter tile puzzle to read Einstein's famous quotation regarding whether random events are part of nature's laws.

I		A	P	L	A	Y		D		D	O	E		I	C	E		I	N	C	S		N	M		C
O	N	V		T		G	T	H	A	E	D		O	D		O	T									

Daily Warm-Ups: Physics

Bohr to Einstein

Albert Einstein believed that purely random events had no place in the explanation of how nature behaves. Niels Bohr believed that the decay of individual radioactive atoms were random events. Einstein refuted Bohr's contention with his famous quotation: "God does not play dice with the universe." Bohr's response became equally famous.

Solve the letter tile puzzle to read Bohr's response.

G O D		T E	U N		H O R S E	D O		U N	W	T
O	R	T H E	L L	N O T	I V E					

155

Atom's Time Line

The structure of the atom that we now learn in school is commonly called the **Bohr model** of the atom, named after Neils Bohr. However, Bohr relied upon the earlier discoveries of others to aid in his final development of how the parts of an atom work together.

Using the list of scientists below, identify which scientist made each of the important discoveries described.

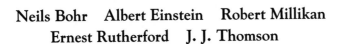

Neils Bohr   Albert Einstein   Robert Millikan
Ernest Rutherford   J. J. Thomson

156

1. Discovered the electron _____

2. Measured the charge of an electron _____

3. Discovered the nucleus _____

4. Theorized the exact orbits of electrons _____

The Curie Family

History has recorded its share of famous families who left their mark in the world of politics, business, and entertainment. Few, if any, can match the recognition achieved in science by the Curie family. Pierre and Marie Curie, husband and wife, are credited with the discovery of radioactivity as an atomic property, as well as the discovery of polonium and radium. Later, their daughter Irene Joliot-Curie and her husband led the way for the production of artificial radioactive elements.

Solve the letter tile puzzle to find out the unique distinction that Marie Curie and Irene Joliot-Curie together hold.

A	ZE	TO	OTH	EL	ER	WIN	AND
D	ATER	EA	UGH	NOB	PRI	Y	M	ONL
CH

157

Set in Stone

The **half-life** is the amount of time that it takes for one-half of a sample's radioactive isotopes to decay. The half-life of carbon-14 is 5,730 years. After 5,730 years, only one-half of the carbon-14 atoms originally in an object are left. After another 5,730 years, only one-half of those remaining (one-fourth of the original isotopes) are left.

In 2003, it was reported that an ancient stone tablet detailing repair plans for the Jewish Temple of King Solomon had been discovered. Questions arose as to whether this tablet was really from King Joash, who lived 2,880 years ago. Among the tests being run to answer that question was a carbon-14 test which showed the tablet to date back to the ninth century B.C.E.

How many half-life periods have passed since the tablet was originally inscribed? Do you think that the tablet is too old for carbon-14 dating to be accurate?

158

Daily Warm-Ups: Physics

A Light Work Day

According to the **special theory of relativity,** an observer on Earth would notice that time passes noticeably more slowly on a spacecraft moving by the earth at a speed near the speed of light.

Suppose your job requires you to travel to nearby planets at speeds close to the speed of light. Your boss pays you for the amount of time you spend traveling. Would you make more money if your time clock stayed on Earth or was located on your spacecraft? Explain your answer.

159

Tachyons

The special theory of relativity sets the speed of light (3×10^8 meters/sec) as the theoretical speed limit for the universe. But what if a particle existed that traveled faster than the speed of light? Could we even detect it? If we used it to communicate, would we receive the message before it was sent? Actually, a name has already been given to this purely hypothetical particle: the **tachyon.**

Read the limerick below about a faster-than-fast woman. Then try your hand at writing a limerick about tachyons.

There once was a woman named Bright
Who could travel much faster than light.
She left one day
In a relative way
And returned on the previous night.

Quark! Quark!

It was once thought that protons, neutrons, and electrons were the fundamental building blocks of matter. We now know that protons and neutrons are made up of even smaller particles called **quarks.** There are six different quarks, each with a peculiar name: *up, down, strange, charmed, bottom,* and *top.*

Find each of the six quarks in the word puzzle.

```
T  T  N  E  D  P  B  Q  B  U
I  O  I  U  E  J  Q  P  O  M
Y  Y  P  X  M  L  D  O  F  G
E  B  Y  U  R  L  V  T  E  M
F  G  G  S  A  X  H  J  O  S
F  R  N  K  H  C  W  T  R  X
G  L  Y  A  C  F  T  M  A  B
D  O  W  N  R  O  V  Y  R  D
R  E  R  Y  B  T  Z  P  U  V
T  F  R  M  J  V  S  P  L  O
```

161

© 2003 J. Weston Walch, Publisher

Atomic Prediction

In 1942, a team of physicists led by Enrico Fermi released the first nuclear chain reaction. The experiment was conducted beneath the bleachers of the football stadium at the University of Chicago. The success of the experiment led to the development of the atomic bomb in 1945. It established the basic principle behind nuclear power plants, which supply a significant portion of the world's electrical energy.

Among those participating in that experiment was Leo Szilard, the Hungarian physicist who first introduced the concept of a chain reaction. After all the cheering and handshaking had died down, Szilard turned to Fermi and said, "This day will go down as a black day in the history of mankind."

In your opinion, did Szilard's prediction prove to be true? Write a paragraph explaining your view.

162

Celestial Objects

The universe is mainly empty space. The study of objects occupying that space, and of their effect on each other, continues to provide answers to questions about our place in the universe.

Find the following seven astronomical terms in the word puzzle.

asteroid   comet   galaxy   moon   planet   quasar   star

```
H O B Z S C J R I C M W Q F U
G P T F U K F M O K N D C S D
M S L A A L M M K I M Q Y I V
G A L A X Y E A D N A U O K L
N O O M N T O P S G Y R F V M
Z G R N F E R R T B E N H Z L
P P M A X P T W H T X J K F T
H H X W T N B Q S D S T W Q X
L J Q P P S L A U E Q C N M Q
B S U J L C C H N V U N H O Q
K X E V D K O R O L A V V O U
R H S Q T N D H U J S A Q A K
X F V Q P E D G K B A R D M A
J R H H U B J V I H R H J D A
A N N G L K T J A C S Q X T H
```

163

© 2003 J. Weston Walch, Publisher

Order! Order! 1

In 1543, Nicolaus Copernicus's *On the Revolutions of the Heavenly Bodies* was published. It was the first work to analytically put the planets in orbit around the sun. Copernicus knew that his work would be very controversial and arranged to have it published upon his death to avoid persecution.

Copernicus's work only included the so-called naked-eye planets, since telescopes were not invented until 1608. Name those planets.

164

Order! Order! II

Copernicus's model of the solar system had been around for more than 60 years before Johannes Kepler gave it the precision it needed to be universally accepted. His three laws of planetary motion made it possible to accurately describe the position of a planet at any point along its orbit.

Daily Warm-Ups: Physics

Driven by his desire to prove the existence of a supreme creator, Kepler attempted to tie the motion of all the planets into one mathematical relationship. His third law provided the relationship he was looking for: $(T^2/r^3) = k$ where T represents the period of the planet's orbit, r represents the average radius of its orbit, and k represents the constant that is always obtained regardless of the planet. Determine the value of k if the planet's distance from the sun were measured in astronomical units and the planet's period of orbit were measured in Earth years. *Hint:* One astronomical unit is defined as the distance from the earth to the sun.

165

© 2003 J. Weston Walch, Publisher

Sky Search

Use the clues to help you in filling in the blanks identifying things found in the sky. *Hint:* Each word begins with **S,** and they are all related to the first word.

1. a self-luminous object

2. the largest type of the object

3. an explosion

4. charged stream

166

1. S __ __ __

2. S __ __ __ __ __ __ __ __ __

3. S __ __ __ __ __ __ __ __

4. S __ __ __ __ __ __ __ __ __

The Long Shortcut

The usual way of getting to a destination in the shortest amount of time is to travel in a path that will result in the shortest distance. However, when trying to get to a destination in outer space, NASA often uses some basic physics principles to get there faster by taking a longer route.

Instead of sending a recent probe directly to Saturn, NASA sent it on a less direct path toward Jupiter, where Jupiter's enormous gravitational pull sped up the probe and changed its direction toward Saturn. Explain exactly how the gravitational force between Jupiter and the probe both sped up the probe and changed its direction without trapping it into orbit around Jupiter.

167

Light Time

Astronomers often speak of how far away certain stars and galaxies are in terms of light-years. But what about measuring the objects in our own solar system? The distance represented by one light-year is many times greater than the size of our entire solar system. Instead of light-years, they often refer to light-seconds, light-minutes, or light-hours.

Using the rounded value of 1 light-year = 1×10^{13} kilometers, calculate the distance from Earth to the solar objects below in the light-units requested.

168

Object	Earth-Object Distance (miles)	Earth-Object Distance (light-units)	
1. Moon	385,000 km	_____	light-seconds
2. Sun	150,000,000 km	_____	light-minutes
3. Pluto	5,740,500,000 km	_____	light-hours

Hubble 11

The Hubble telescope's primary mirror has a diameter of 8 feet. This is not particularly large when compared to the largest telescopes here on Earth.

The largest ground-based telescopes have mirrors that are four times the size of the Hubble telescope. Explain why the Hubble can nonetheless see much farther, and with much more detail, than any of the ground-based telescopes.

169

Looking Back

In 1994, astronauts made a correction to the Hubble telescope's mirror assembly that allowed us to see farther than we had ever seen before. One of the first images we obtained was from the central regions of the galaxy M100 in the constellation Virgo, some 56 million light-years away.

As we viewed those images, we were seeing that region "live" 56 million years ago. Suppose intelligent life was in that region and was looking through their own telescope at Earth. Also imagine that their telescope could see details of our planet including any life forms inhabiting the planet. Describe what they might see.

170

Shrinking Star

When a star dies, its fusion furnace goes out, and gravity is left to have its way. The dead star will collapse upon itself according to how much mass it has. The seemingly odd part is that the bigger the star is, the smaller it will become in diameter, even though it will contain more mass. Of course, if you consider that the amount of gravitational force varies directly with the amount of mass, this is not really so odd after all.

Our sun, an average-sized star of diameter 1,400,000 kilometers, will collapse to become a white dwarf of diameter of 12,000 kilometers (about the size of the earth). A star that is about 1.5 times the diameter of our sun will collapse to become a neutron star with a diameter of roughly 24 kilometers (the size of Manhattan Island). A star that has a diameter of 10 times our sun will probably become a black hole. Estimate what you think its diameter might be. (Note: The diameter of a black hole is not a real dimension but is a measure of the high-gravity area around the black hole.)

171

Death Watch

All stars go through a similar cycle of birth, maturity, and death. How long they last as an actual star and what they become upon their death is determined by their mass. Our sun is an average-size star, significantly bigger than the nearby star Proxima Centauri, and many times smaller than the star Betelgeuse.

For each of these three stars, draw a line connecting the correct estimate for its life span as a star and a line connecting the correct final form it will take upon its death.

Time-span	Star	Final Form
millions of years	our Sun	black hole
billions of years	Proxima Centauri	white dwarf
trillions of years	Betelgeuse	brown dwarf

Lost and Found

Each and every second, our Sun's core continues to fuse 600 million tons of hydrogen into 595 million tons of helium. Using Einstein's famous equation, $E = mc^2$, we find that the missing 5 million tons of matter has been converted into an amount of energy equal to 1 billion one-megaton hydrogen bombs!

Because the sun radiates its energy outward in all directions, and because the Earth is 93,000,000 miles away, the Earth only receives one-half of one-billionth of the Sun's energy. You might ask how we can survive with such a small percentage. Calculate how much energy that would be in one hour in terms of one-megaton hydrogen bombs.

173

Black Holes

A **black hole** is really not a hole at all, but rather the incredibly dense matter that resulted when a giant star reached the end of its life cycle. The ability of a black hole to exert a strong gravitational force far exceeds what it could exert when it was still a star.

Suppose a distant planet was in orbit around the star. Describe how the orbit of the planet would be affected if the star became a black hole.

174

Finding the Mass of a Black Hole

Although the existence of black holes was predicted as far back as Einstein's time, it has only been in fairly recent times that astronomers have found them. What they have actually seen is a companion star that is orbiting inward toward the black hole. The measurements they have been able to make of the companion star include the radius of its orbit around the black hole and the amount of Doppler shift observed as it moves in that orbit.

Using what you know about Doppler shift, circular orbits, and the universal law of gravity, explain how you could use the astronomers' measurements to determine the mass of the black hole.

175

Dividing by Zero

A **black hole** is defined as a region of space-time from which nothing, not even light, can escape because gravity is so strong. It is a concept that fascinates people of all ages and is the subject of countless questions, especially ones about what happens if you enter a black hole. Most of these questions have no answers because Einstein's theory of relativity predicts the breakdown of the physical laws of nature themselves as you reach its boundary.

Without physical laws present, there is no basis for making any prediction as to what would happen in any type of event. Comedian Steven Wright summed it up with a well-known quotation. Solve the letter tile puzzle to find out what he said.

176

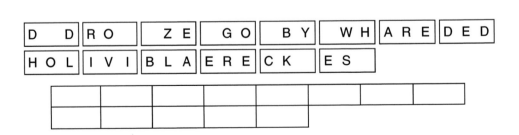

| D | DRO | ZE | GO | BY | WH | ARE | DED |
| HOL | IVI | BLA | ERE | CK | ES |

Why Scientists Need to Know How to Write

Most science students are familiar with the ideas, concepts, and laws that have shaped our view and understanding of the universe. These were perhaps learned by studying a textbook or listening to a teacher. But how did the scientists who originally developed these ideas get anyone to understand what they were thinking? The answer to that question lies in their writing skills. By writing a book or paper that clearly describes the problem they are addressing and then logically presents their rationale for their answer, scientists can advance the forefront of knowledge.

Isaac Newton, Nicolaus Copernicus, Galileo Galilei, and Albert Einstein were all famous scientists who changed our view of the universe and the laws that govern it through famous writings. Use what you know about their work to match each scientist with the title of his publication.

Title
1. *On the Revolutions of the Heavenly Spheres*
2. *The Principia*
3. "On the Electrodynamics of Moving Bodies"
4. *Discourses on Two New Sciences*

Scientist
a. Albert Einstein
b. Galileo Galilei
c. Isaac Newton
d. Nicolaus Copernicus

Job Search

Certain careers, such as being a research physicist or astronomer, obviously deal with physics. The connection to many other careers may not be as obvious to many people. The study and application of particular areas in physics are critical in many different jobs.

For each job title, choose the area of physics that applies the most and write it in the blank to the right. Some jobs may involve more than one area.

Areas of Physics		
mechanics	thermodynamics	electricity and magnetism
atomic and nuclear	waves and optics	fluid dynamics

178

Job Title	Area(s) of Physics
1. radiologist	_____
2. civil engineer	_____
3. meteorologist	_____
4. seismologist	_____
5. oceanographer	_____
6. archeologist	_____
7. photographer	_____

A Clean Start

In 2003, a company in Germany created an interactive washing machine. It understands spoken-word commands, such as, "Pre-wash, then hot wash at 95 degrees, then spin at 1,400 revolutions." It can also give verbal instructions on how to use the machine! Of course, this all comes with a price.

The inventors of the interactive washing machine have been quoted as saying that it is needed because electronic appliances have become too complicated for consumers to use. However, some critics have suggested that this may be a misplaced attempt to aid the technologically challenged consumer. Choose a different appliance in which a similar voice-recognition system would provide a greater benefit to its user and explain how it would help.

179

Changes

Advances in modern technology have improved our methods and ability to communicate to levels unthinkable just a few decades ago. Much of this technology relies on communications satellites put into very high orbits. The smaller the satellite and the higher the orbit, the longer the satellite will stay in orbit. It has been predicted that some of these current satellites will remain in orbit for millions of years.

Obviously, these satellites will outlive their usefulness, as the technology that once required them will drastically change. But what about changes to Earth itself? Describe a few of the changes that will occur to the planet by the time the satellite comes out of orbit.

180

1. Answers will vary.
2. Answers will vary—e.g., the big bang or theory of evolution vs. creationism.
3–6. Answers will vary.
7. 1. a tape measure
   2. a kitchen timer
   3. electronic mass balance
   4. a caliper
8. 1. 10^{-2} 4. 10^1
   2. 10^{-1} 5. 10^0
   3. 10^1
9. 1. kilogram   5. watt
   2. meter, mile   6. second
   3. joule, horsepower 7. joule, horsepower
   4. ounce, pound   8. ounce, pound
10.

Quantity	"Famous Unit"	Contribution
1. force	newton	laws of motion and gravity; color
2. work, energy	joule	concepts of work and energy
3. power	watt	steam engine
4. electric charge	coulomb	electrical force

5. radioactivity   curie   discovery of radio-active atoms

11. 1. $kg \cdot m/s^2$
    2. $kg \cdot m^2/s^2$
    3. $kg/m \cdot s^2$
    4. $kg/m \cdot s^3$
12. 1. c   2. e   3. b   4. a   5. d
13. 1. 42 days (1,000 hrs)
    2. 81 days (1,948 hrs)
    3. 16 years (142,680 hrs)
    4. 117,728 years 1,032,000,000 hrs)
14. 2.3 million years
15.

G L O	B A L		P O	S I T	I O N	I N G		S Y	S T E
M									

16. 1. instantaneous   3. instantaneous
    2. average   4. instantaneous
17. She cannot do it. In order to average 100 mph for the entire 100 miles, the time must be exactly 1 hour. She has already used up 1 hour at the halfway mark.

Daily Warm-Ups: Physics

18.

```
P Z T M M M B O O F E Z
E T H E R M A S C O P E
N G I Z L M X E O J X A
D U W A T E R C L O C K
U Q R C P H S Q S T K X
L M A H T O P C Q M W H
U X B N M P I V O I Q P
M H K R Q P L W W P O A
U C E S Z S B Z F K E K
```

19.

Description of Motion	Direction of Velocity	Direction of Acceleration
A ball is dropped from a ladder.	↓	↓
A car is moving to the right when the driver applies the brakes to slow down.	→	←
A ball tied to a string and being swung clockwise is at the top of its circular path.	→	↓
A sled is pushed to the left causing it to speed up.	←	←

20. Answers will vary. E.g., go due east 40 m, turn and go 50° N of E 30 m, turn and go due East 70 m.

21.

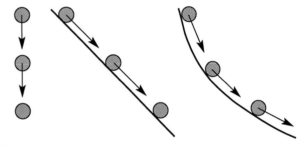

22.

Newton's Law of Motion	Example
First law	Answers will vary. E.g., a car sitting still at red light
Second law	Answers will vary. E.g., a baseball bat hitting baseball
Third law	Answers will vary. E.g., a fist slowing down as it hits punching bag

23. Light would only need 0.00005 seconds to make the round trip.
24. His diameter was 126,720 km; it was important because it showed that any timer used to measure the speed of light would have to be capable of measuring much smaller increments of time than were currently possible.
25. 80.5 km apart
26. 1. ✓
    2. ✓
    3. The plane's motor is exerting a force.
    4. ✓
    5. The pressurized water is exerting a force.
27. 36.5 m/s
28. Longer distances require a lower launch angle (45° is the maximum). The lower the launch angle, the lower the ball as it crosses the line of scrimmage.
29. 1. tangential   3. tangential
    2. angular   4. translational
30. 2.67 rev/min.
31. Seventy-eight rpm is larger because the faster speed requires a longer groove to finish in same amount of time.

32. Twenty-four hours. This period is important in that a communications satellite needs to remain overhead.
33. **Order of Altitude**
    (lowest)   Boeing 747 airplane
               Freedom 7 Mercury space capsule
               Space shuttle
               International Space Station (ISS)
    (highest) Hubble space telescope
34. • light = 1.27 seconds; spacecraft = 9.44 hours
    • light = 0.04 seconds; spacecraft = 56.5 minutes
35. 93 minutes
36. Isaac Newton
37. The force from behind sends the car forward. The person's body is sent forward as well by the seat exerting a forward force. The head has nothing pushing it forward and therefore remains still while the rest of the body goes forward causing the strain on the neck.
38. 1. to the left   4. (no net force)
    2. inward   5. upward
    3. inward

39.

Reaction Force	Reaction Effect
golf ball hits golf club	golf club slows down
ball pulls earth up	earth accelerates upward
earth pulls outward on sun	sun wobbles in its path
friend's chest pushes against boy's hands	boy's hands slow down

40. Answers will vary.

41.

42. 1. The water pushes *forward against the fish*.
2. The road pushes *forward against the tires*.
3. The water pushes *forward against the swimmer*.
4. The floor pushes *forward against your foot*.

43.

OBJ	ECT	S W	ILL		RE	MAI	N	A	T	R
EST	OR	IN	UN	IFO	RM	MOT	ION			
UN	TIL	AC	TED	UP	ON	BY	A N			
ET	FOR	CE								

44.

STA	NDI	NG	ON	THE		SH	OUL	DER
S	OF	GIA	NTS					

45.

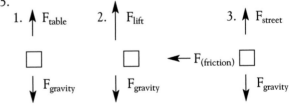

Daily Warm-Ups: Physics

46.

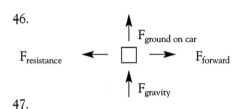

47.

Mass (kg)	Weight (N)
1	9.8
0.1	1
25	245
2.6	25

48. It would decrease. You would have mass below you pulling downward and mass above you pulling upward. At the center of the earth, you would weigh zero.

49.

WHA	T G	OES	UP	MU	ST	COM	E D

OWN

50. The hammer and feather hit the ground at the same time. On Earth, the hammer would hit first due to the effects of air friction on the feather.

51. 1. 490 newtons   3. 1,245 newtons
    2. 80 newtons

52. Answers could vary; they should all contain less acceleration and slower speeds.

53. Answers will vary due to the individual's weight. A 150-lb person will have a mass of 4.7 slugs and a weight of 668 newtons.

54.

55. The higher center of gravity on the SUV is not within the support area defined by the tires when on the banked curve.

56. 1. low
    2. high
    3. lower than hard
    4. higher than soft

Daily Warm-Ups: Physics

57. Answers will vary
58.

Situation	Object exerting the centripetal force
A ball is tied to a string and swung in a horizontal circle.	the string
The clothes and water inside a washing machine are spun in a circle during the spin cycle.	the wall of the tub
A car traveling on the road follows a curve in the road.	the road
A marble is rolled around the inside edge of a paper plate.	the inside edge of the plate
A penny rolls on its edge on the inside of a balloon.	the inside of the balloon

59. Both the clothes and the water have velocities tangent to the circular path they are traveling in. The wall of the tub pushes inward except where there are holes. The water escapes in a path tangent to the circle and the clothes are compressed tangentially against the wall.

60. It will stay at rim level. Ice displaces an amount of water equal to its weight. When melted it will take up the same amount of space as the water it displaced.
61. The water goes down. Since balls are more dense than water, they displace more volume while floating than when submerged.
62. Regular cola uses regular sugar and the density of the can of cola is slightly greater than water. Diet cola uses a sugar substitute which is sweeter and therefore less is used. The density of the can of diet cola is slightly less than water.
63. 21,854,000 gallons
64. The density of warmer sea water is lower, causing the ship to sink lower. Fresh water also has a lower density than sea water, causing it to sink even lower.
65. Area = 37.5 in²; cooler air results in lower air pressure in tires, thus more area is in contact with the ground (flatter).
66. Measure the length and width of each "footprint" to determine its area. Multiply the area times the pressure to get the number of pounds being supported. Add the results for the four tires.

67. **North Pole**   **Equator**
    1. sink   rise
    2. cold air contracts   hot air expands
    3. high   low

68. F (buoyant) = pressure change × area = 7.35 lb/in²
    × 14,400 in² = 105,840 lbs.

69. The water pressure is 23 times greater at the bottom of the dam.

70. The moving water causes the air inside the shower to move. Air inside is now exerting less pressure outward on the curtain than the stationary air on the outside pushing in.

71. Doors open by rotating at the hinges. Knobs at the end mean greater distance/less force. Big knobs mean a longer turning radius/less force.

72. (easiest to rotate)   solid ball 15 cm diameter
                          solid ball 30 cm diameter
    (hardest to rotate)   hollow ball 30 cm diameter

73.
```
U J C Y Z A M T Q C
U H I E Q N K V H F
L A C N Q F L A W J
A G C D U Z R U L T
T D A O V G J A Q V
E Q C E E S V Y O J
A A R N L Q D I L B
N S S A M O U Q G A
F T T G P Q P H Z M
B D P M R I Z T V J
```

74. electromagnetic

75. #4. The Coanda effect channels the water back into the gutter while providing no opening for leaves or debris.

Daily Warm-Ups: Physics

76.

Opinion	Jobs	Physics
(hardest) Answers will vary.	store clerk	(most) furniture mover
	accountant	package delivery driver
	package delivery driver	store clerk
(easiest)	furniture mover	(least) accountant

77.

Device	Energy In	Energy Out
toaster	electrical	heat (light)
portable generator	chemical	electrical
electric dryer	electrical	heat
flashlight	electrical	light (heat)

78. Stopping Distance Rank
truck moving at 15 mph
car moving at 30 mph
car moving at 60 mph
car moving at 75 mph
truck moving at 60 mph

79. $11.50

80.

```
S A A P T K D X C S
B L M J O K I L A W
R O P S K L L G K W
G C Z U A V O P Y R
D I U Q I L S C N A
E W W G V G P A E J
C F M W I P L B S Q
B D Y Z B J B L O N
X F N C B K B Z W Y
```

81. (hottest)   100 K
                 212° F
                 96° C
                 0° C
                 0° F
  (coldest)   0 K

82. The rim of the lid is a circle. As the metal expands, the circumference expands. Metal expands more than glass.

83. 1. 506° C   3. 620°
    2. 546° C   4. 746°

84. Both are at same temperature (room temperature), but the porcelain tile conducts heat away from your feet faster.
85. While the more even heating is somewhat beneficial, the fact that it is far less likely to shatter is the main reason it is better.
86. The silver lining prevents loss due to radiation and the vacuum prevents loss due to conduction.
87. Being at a higher temperature, the block transferred heat to the meat by conduction. As the block cooled down, the rate of conduction slowed. Hot water raised the temperature of the block.
88. Styrofoam is mainly air. Heat travels through it poorly; it can't leave the hot coffee or get to the cold drink.
89. Death Valley: Use the thermometer to regulate temperature at 100° C, even though it is not boiling. Denver: Use the pressure cooker to raise the pressure to 1 atmosphere.
90. Northern U.S.—darker, due to longer winters, shorter summers, and less direct sunshine

Southern U.S.—lighter, due to longer summers, shorter winters, and more direct sunlight
Central U.S—either, as it would not make much difference either way
91. The time to bring a person to a stop is increased, thereby decreasing the force exerted on the person by the belt.
92. Answers will vary, but may include airbags, suspension bumpers, and padded dashboards.
93. Impulse. The longer stopping time after contact means smaller force.
Not as much protection—unlike in a car, the rider would continue to move.
94. Conservation of momentum
95. Going into a tuck decreases the moment of inertia and increases the rate of spin; coming out of the tuck increases the moment of inertia and decreases the rate of spin.
96. 34.1 m/s
97. Answers will vary, but may include pool balls on a pool table or bowling balls in the ball return rack.

98. The law of conservation of energy and the first law of thermodynamics do not allow for the creation of energy.

99. Friction between the parts would convert some of the energy to heat, which could not be totally retrieved.

100. Answers will vary, but may include walking across carpet or clothes tumbling in the dryer.

101. Moisture drains the static charge.

102.

Electric Force Present?	Repel or Attract?
yes	repel
no	
no	
no	
yes	repel
yes	attract

103.

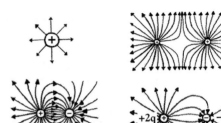

104.

```
      h E a t
 s  o  u N d
      m E c h a n i c a l
        R
    l i G h t
        Y
```

105. In the series circuit, the 60-W bulb burns brighter because the 75-W bulb has higher resistance and gets less voltage. In the parallel circuit, the 75-W bulb burns brighter because the power is greater.

106. Five can safely be used. Each cooker draws 7.5 amps.
107. Each appliance needs it own path so that it does not affect the operation of other appliances. As you add more appliances to a parallel circuit, the total current increases.
108. Although a bit unlikely, the idea is that by jarring the domains in the presence of the earth's magnetic field, they will align with it.
109. 1. no current in the wire
     2. electron is not moving
     3. X
     4. X
110. 1. X
     2. no current in the wire
     3. X
     4. X
111. It is more like an electromagnet, because the movements of the metal liquids create electrical currents which change in strength and direction.
112. Answers will vary, but may include such things as heart pacemakers stopping and pigeons becoming disoriented.

113. Answers will vary. Possible answers are shown.

Maxwell's Equation #	Example or Occurrence
1. Unlike charges attract each other; like charges repel.	static cling
2. Magnetic poles cannot be isolated from each other.	bar magnet
3. Electric currents create magnetic fields.	electromagnet
4. Changing magnetic fields can create electric current.	electric generator

114.

Wave	Source
A tuning fork is struck with a rubber hammer, producing a sound wave.	the vibrating tuning fork
A motorboat moves through the water, leaving its wake behind.	the propeller blades
A performer sings a high note.	vocal chords
A light bulb gives off light.	vibrating electrons

115. wave**electromagnetic**sound**lanidutignol**ight**mechani-cal**igh**transverse**energy

116.

Wave	Type
a sound wave	longitudinal
a water wave caused by a boat moving	transverse
a wave in a rope caused by one end being moved up and down	transverse
a wave in a coiled spring caused by pushing one end in and out repeatedly	longitudinal
a light wave	transverse

117. Answers will vary, but may include the breaking of glass by a singer.
118. The radio transmission travels much farther but it travels at the speed of light. The actual sound wave travels through the ballpark air at the speed of sound.
119. distance = 1 mile, t_{light} = 1/300,000,000 second (instantaneous), t_{sound} = 4.7 seconds in one mile
120. f = 142 Hz

121. Answers will vary.
122. light
123.

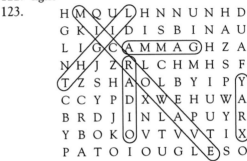

124.

RAD	IO	DET	ECT	ION	AN	D	R	ANG
ING								

125. AM: λ = 322.6 m   FM: λ = 2.8 m
     AM waves are too long to make it through underpasses unaffected while FM waves are not.
126. time = 3,200,000/300,000,000 m/s = 0.01 sec
     Therefore, the accuracy of one second is not affected.

127. When the light waves strike the transparent material, a chain of absorptions and reemissions occur through the material. The time delay between each absorption and reemission produces an average speed of light less than 3×10^8 meters per second.

128. Answers will vary, but may include infrared radiation (feeling the warmth of the sun) and ultraviolet radiation (sunburn).

129.

		L	I	G	H	T			
A	M	P	L	I	F	I	C	A	T	I	O	N
				B	Y				
	S	T	I	M	U	L	A	T	E	D
	E	M	I	S	S	I	O	N		O	F
	R	A	D	I	A	T	I	O	N

130. The red and blue parts are absorbed; green is reflected, as evidenced by the green leaves.

131.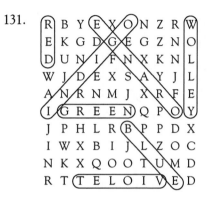

132. The prefixes relate to energy. Infrared has less energy than red and ultraviolet has more energy than violet.

133. 1. green
2. red
3. blue

194

134.

```
N Y X E E F S Q J R
K E O W O O X R T Q
A M E H N D Z J E S
P V K R Y Z T O U Z
S C D U G V D K L S
U V M E L E F Q B B
B U Y H R Y M E C Q
K N Y P C Z F Y Z V
```

135. At night, you are seeing transmitted light from the outside but only reflected light from the inside.

136. absorption
reflection

137. Answers will vary, but may include:
reflection   any opaque object
refraction   any transparent object

138.

```
F N D C K Z W T J O N R
A Y T Q K L Z G W O D E
S S T B Q J Y D I G V F
R T H K E Z Z T Q U L L
W H J M Q U C F O R X E
N O I T C A R F E R C C
A B S O R P T I O N R T
I J U F P R E C P K G I
F N F B D G O O D D S O
C D I F F R A C T I O N
```

139. 1. virtual   3. virtual
2. virtual   4. real

140. The image is real. It is actually upside down with respect to the film, which itself is upside down in the projector.

141. A camera can use either reflection using mirrors or refraction using a diverging lens to produce a virtual image of the object being photographed.

142. The mirror would have to be at least half your height.
143. Answers will vary, but may include tot, toot, etc.
144. The plane mirror is best because it gives the true distance between you and the car behind you.
145. Since the light coming from the water to your eye is refracted away from the normal, the container appears to be farther away than it really is. You should drop the coin in front of the container.
146. One side of the lens is more curved than the other side.
147. As a person ages, the eye muscles around the lens become less flexible, decreasing their ability to contract.
148. Nearsighted. Eyes of nearsighted people are slightly longer than usual from front to back, so the retina is located farther back in the eye. Under water, there is less refraction than in air. This causes images to form farther back in the eye, closer to the retina.

149.

$$\underset{23}{I}\ \underset{17}{M}\ \underset{1}{A}\ \underset{9}{G}\ \underset{23}{I}\ \underset{21}{N}\ \underset{1}{A}\ \underset{5}{T}\ \underset{23}{I}\ \underset{3}{O}\ \underset{21}{N}\quad \underset{23}{I}\ \underset{20}{S}\quad \underset{17}{M}\ \underset{3}{O}\ \underset{18}{R}\ \underset{4}{E}$$

$$\underset{23}{I}\ \underset{17}{M}\ \underset{6}{P}\ \underset{3}{O}\ \underset{18}{R}\ \underset{5}{T}\ \underset{1}{A}\ \underset{21}{N}\ \underset{5}{T}\quad \underset{5}{T}\ \underset{11}{H}\ \underset{1}{A}\ \underset{21}{N}$$

$$\underset{19}{K}\ \underset{21}{N}\ \underset{3}{O}\ \underset{7}{W}\ \underset{14}{L}\ \underset{4}{E}\ \underset{16}{D}\ \underset{9}{G}\ \underset{4}{E}$$

150.

THE	SP	EED	OF	LI	GHT	IS	CO
NST	ANT	FO	R A	LL	OBS	ERV	ERS

151. Answers will vary.

152.

KEE	PIN	G Y	OUR	MO	UTH	SH	UT

153. Albert Einstein

154.

I	A	M C	ONV	INC	ED	THA	T G	OD
DOE	S N	OT	PLA	Y D	ICE			

Daily Warm-Ups: Physics

155.

DO	NOT	TE	LL	GOD	HO	W	T	O	R
UN	THE	UN	IVE	RSE					

156. 1. Thomson
2. Millikan
3. Rutherford
4. Bohr

157.

ONL	Y	M	OTH	ER	AND	DA	UGH	TER
TO	EA	CH	WIN	A	NOB	EL	PRI	
ZE								

158. Almost one-half of one half-life has passed.
No, there are still many carbon-14 isotopes left to count.

159. More time will pass on the earth clock.

160. Answers will vary.

161.

```
T T N E D P B Q B U
I O I U E J Q P O M
Y Y P X M L D O F G
E B Y U R L V T E M
F G G S A X H J O S
F R N K H C W T R X
G L Y A C F T M A B
D O W N R O V Y R D
R E R Y B T Z P U V
T F R M J V S P L O
```

162. Answers will vary.

163.
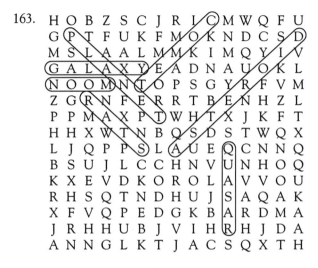

164. Mercury, Venus, Earth, Mars, Jupiter, Saturn

165. $1 \text{ yr}^2/\text{A.U.}^3$

166. 1. STAR
   2. SUPERGIANT
   3. SUPERNOVA
   4. STELLARWIND

167. The path that the probe was on was not perpendicular to the gravitational pull. Therefore, one component of the force acted perpendicular as a centripetal force (changing its direction), while another component acted in the direction of the probe's motion (speeding it up).

168. 1. 1.2 light-seconds
   2. 7.9 light-minutes
   3. 5 light-hours

169. The light does not have to come through Earth's atmosphere.

170. They would see Earth 56 million years ago, shortly after the age of dinosaurs.

171. Only a few miles

172.

Time-span	Star	Final Form
millions of years	our sun	black hole
billions of years	Proxima Centauri	white dwarf
trillions of years	Betelgeuse	brown dwarf

173. It would be the equivalent of 1,800 one-megaton hydrogen bombs each hour.

Daily Warm-Ups: Physics

174. It would be unchanged. The mass did not change appreciably and the distance between the star and the planet did not change.
175. Use the Doppler shift to determine the speed (v) of the companion star. The speed and radius of orbit can then be used to find the centripetal force, which is the gravitational force. The gravitational force can then be used to find the mass.
176.

BLA	CK	HOL	ES	ARE	WH	ERE	GO
D D	IVI	DED	BY	ZE	RO		

177. 1. d 2 c 3. a 4. b
178. 1. atomic and nuclear
    2. mechanics
    3. thermodynamics
    4. waves and optics
    5. fluid dynamics, electricity and magnetism
    6. atomic and nuclear
    7. waves and optics
179. Answers will vary.

180. Answers will vary, but may include the continents moving apart due to plate tectonics.